D1352254

00319453

Titles of Related Interest

ANNALS OF THE ICRP, Volume 14/2

Protection of the Public in the Event of Major Radiation Accidents: Principles of Planning

ANNALS OF THE ICRP, Volume 15/1

Principles of Monitoring for the Radiation Protection of the Population

ANNALS OF THE ICRP, Volume 15/2

Protection of the Patient in Radiation Therapy

Journals of Related Interest

HEALTH PHYSICS: The Radiation Protection Journal

INTERNATIONAL JOURNAL OF RADIATION ONCOLOGY · BIOLOGY · PHYSICS

CHERNOBYL

The Real Story

by

RICHARD F. MOULD

MSc. PhD, CPhys, FInstP, FIS, FIMA

PERGAMON PRESS

OXFORD · NEW YORK · BEIJING · FRANKFURT

SÃO PAULO · SYDNEY · TOKYO · TORONTO

U.K.	Pergamon Press, Headington Hill Hall, Oxford OX3 0BW, England
U.S.A.	Pergamon Press, Maxwell House, Fairview Park, Elmsford, New York 10523, U.S.A.
PEOPLE'S REPUBLIC OF CHINA	Pergamon Press, Room 4037, Qianmen Hotel, Beijing, People's Republic of China
FEDERAL REPUBLIC OF GERMANY	Pergamon Press, Hammerweg 6, D-6242 Kronberg, Federal Republic of Germany
BRAZIL	Pergamon Editora, Rua Eça de Queiros, 346, CEP 04011, Paraiso, São Paulo, Brazil
AUSTRALIA	Pergamon Press Australia, P.O. Box 544, Potts Point, N.S.W. 2011, Australia
JAPAN	Pergamon Press, 8th Floor, Matsuoka Central Building, 1-7-1 Nishishinjuku, Shinjuku-ku, Tokyo 160, Japan
CANADA	Pergamon Press Canada, Suite No. 271, 253 College Street, Toronto, Ontario, Canada M5T 1R5

First edition 1988

Library of Congress Cataloging in Publication Data

Mould, Richard F. (Richard Francis)
Chernobyl—the real story.
Bibliography: p.
Includes index.
1. Chernobyl Nuclear Accident, Chernobyl,
Ukraine, 1986. 2. Chernobyl Nuclear Accident,
Chernobyl, Ukraine, 1986—Environmental aspects.
I. Title.
TK1362.S65M68 1988 363.1'79 87–32855

British Library Cataloguing in Publication Data

Mould, R. F.
Chernobyl: the real story.
1. Chernobyl Nuclear Accident, Chernobyl;
Ukraine, 1986
I. Title
363.1'79 TK1362.S65C4

ISBN 0-08-035718-0 Hardcase
ISBN 0-08-035719-9 Flexicover

Printed in Great Britain by A. Wheaton & Co. Ltd., Exeter

TO MY DAUGHTER FIONA

Contents

Preface

"Now, what I want is Facts".
Thomas Gradgrind, in *Hard Times*
(Charles Dickens, 1812–70)

MY OBJECTIVE in writing this book was to produce an historical account of what happened before, during and after the accident, rather than embroiling myself in the political arena over Chernobyl's implication for the environmentalist issue. I am sure that volumes will be written on the possible consequences of the catastrophic accident on 26 April 1986 for civil nuclear power programmes. However, I believe a detailed chronicle of Chernobyl is just as important in its own right. My interests led me to research the events of the accident and particularly to try and obtain as much photographic evidence as possible. This is because photographs enable a reader to visually appreciate the actuality of the tragedy and its aftermath, albeit at second hand.

The nine chapters are arranged in a logical order with the nuclear power plant described first, followed by a chapter on the city of Kiev. This is included to emphasise that however horrific the initial toll of the disaster, its effects might easily have extended more dramatically to the Soviet Union's third largest city. The accident and its causes are described in Chapter 3. The fate of the immediate victims of the first 4 months, mainly the firemen who fought the blaze at great personal risk, is given in Chapter 4. Some pictures in this chapter will be distressing, but the realities of Chernobyl should not be covered up.

Chapters 5 and 6 detail the massive operation undertaken, at remarkable speed by the Soviet authorities, to evacuate 135,000 people and several thousand head of cattle; to provide medical attention; to build new towns to rehouse the evacuees; to decontaminate not only the immediate area surrounding the power plant, but also the environment within the 30-kilometre evacuation zone, including agricultural land and forests. Chapter 7 considers the effects of radioactive contamination on the food chain, particularly milk and leafy vegetables, by iodine-131 and caesium-137 and covers not only the Soviet Union but also thirty-four other countries. Chapter 8 describes the entombment of the stricken reactor, the completion of which was reported in *Pravda* on 15 November 1986. Chapter 9 is devoted to the follow-up to the disaster and includes recommendations by the International Atomic Energy

Agency. This chapter also contains estimates of excess cancer deaths (see also Appendix 3) from various sources. A Glossary is provided and a comprehensive list of References.

The photographs, with captions, are placed in chronological order in central sections. This arrangement enabled, without difficulty, the addition of the most recent photographs as they became available prior to the final publishing production deadline of 11 November 1987. These sections therefore present a unique visual historical record spanning the first 18 months following the accident.

Any book of this nature owes a debt of gratitude to many people and organisations, not least to my friends and colleagues in the International Atomic Energy Agency in Vienna and in the World Health Organisation in Geneva; to many picture agency librarians, particularly Mrs Lyudmila Pakhomova from TASS in London, who also made available all the relevant press cuttings from *Pravda* and *Izvestia* and provided colour transparencies for lecture slides of many of the black and white illustrations; and to press agency librarians, especially Mr Ralph Gibson from Novosti in London, who provided me with all Chernobyl press releases sent from their Moscow office.

In order to canvas international opinion and obtain useful material I have spoken to or corresponded with friends who are nationals of Austria, Czechoslovakia, Egypt, France, Federal Republic of Germany, Greece, Hungary, India, The Netherlands, Poland, Romania, Sweden, United Kingdom, United States and the Soviet Union. I am most grateful for their contributions, which often took the form of published or unpublished material relating to analyses of events at Chernobyl. Officials of organisations such as the Soviet Permanent Mission in Geneva, the National Radiological Protection Board in the United Kingdom and the American Food and Drug Agency and National Institutes of Health in Bethesda have also been very helpful.

I have drawn heavily on the working documents provided by the Soviet delegation at the Post-Accident Review meeting, 25–29 August 1986, at the IAEA in Vienna (see Appendix 1), and on 13 hours of tape recordings I made quite openly at this meeting, where I was fortunate to have been a member of the United Kingdom delegation. I would like to emphasise, however, that this book in no way represents any "official" United Kingdom viewpoint, but is only a personal attempt at objectivity, emphasising facts rather than opinions and following my earlier well-illustrated historical book entitled *A History of X-rays and Radium*, which was published in 1980.

Other major documentary sources in addition to the Soviet delegation reports have been papers relating to the two meetings on Chernobyl sponsored by the World Health Organisation, held in May 1986 in Copenhagen and in June 1986 in Bilthoven; articles in scientific journals such as *Nature* and in more popular magazines such as *New Scientist*, *Time* and *Newsweek*; newspapers, mainly but not exclusively from the United Kingdom; videotapes from British television programmes reviewing the accident; Soviet publications such as *Pravda*, *Izvestia* and *Moscow News*; reports from TASS and Novosti; and publications from the Central Electricity Generating Board and the National Radiological Protection Board of the United Kingdom. The data given in Chapter 9 is from visits to Chernobyl by Dr Hans Blix, Director General of the IAEA (May 1986 and January 1987); by Mr Peter

Walker, Minister of Energy in the British Government (December 1986); by a delegation from the General, Municipal, Boilermakers & Allied Trades Union (December 1986); by two small groups of Western journalists (December 1986 and June 1987); and most recently by a three-man group from the Central Electricity Generating Board (October 1987).

I must now single out four persons. I am grateful to Lord Marshall, Chairman of the CEGB, for his encouragement of my embryo ideas for this book and also for his permission to use material from CEGB publications. To Academician Leonid Ilyin, who generously gave me his time to discuss this book during the 25–29 August 1986 meeting, I owe a debt of gratitude for his tacit approval of my draft suggestions of what he termed a "popular-scientific book". Professor Ilyin is Vice-President of the Soviet Academy of Medical Sciences and was in overall charge of the organisation of the medical response following the accident. To Lord Ennals for his support and his agreement to chair the press launch at the House of Lords, Westminster. Finally, I would like to thank my historian daughter, Fiona, to whom this book is dedicated, for her encouragement and her dedicated practical help in transcribing the 13 hours of tape recordings from the Post-Accident Review meeting in Vienna.

RICHARD F. MOULD

11 November 1987
London

Afterword: To Chernobyl in 48 hours

At 1100 hours on 30 November I was asked by Novosti Press Agency in London "Can you be on the night train to Kiev tomorrow?" Almost 48 hours later, to the minute, I was crossing the boundary of the 30-kilometre zone at 100 km/hour in a minibus, after travelling from London by Aeroflot and the 858 kilometres from Moscow to Kiev on the Ukraine night express, train number 1. The organisation and scheduling for my trip, the 183rd official delegation to visit Chernobyl, was truly impressive. I was to spend a total of 6 hours within the 30-kilometre zone with a programme tailored to my specific requests which were based on what I had already learned whilst writing and researching *Chernobyl—The Real Story*. In addition, all my questions on the accident and its follow-up were answered without any problems, and no photographs were prohibited anywhere, clearly demonstrating "Chernobyl-glasnost". Also, I was privileged to be treated as a one-man official delegation and to meet and to be accompanied to the control room of unit No. 1 and to the turbine hall by Mr Mikhail Umanets, Director of the Power Station. I was also pleased, whilst in the control room, to be able to present my opinions to Ukranian TV on the success of the enormous tasks involved in solving the problems caused by the accident, such as the evacuation and rehousing of a population of more than 100,000 persons, building the sarcophagus, complicated decontamination procedures (some work in Pripyat which was expected to take 20 years has taken only 1 year), and, on the very day of my visit, 2 December, the start-up of unit No. 3 by once again operating turbine No. 5. Touring the vast power station complex of buildings and its even vaster surrounding areas, it is still difficult to really appreciate, particularly when facing a snow-capped sarcophagus, the total effort in manpower, in technical expertise, for building materials such as concrete and metal structures, for transport and for financial resources (at least the equivalent of £2 billion sterling) which was required for mobilisation in what was, relatively speaking, a short space of time.

Suppositions about the effects of radiation were also discussed, such as Dr Robert

Gale's estimate of 75,000 cancer deaths due to the Chernobyl accident. However, it was the surprisingly widespread belief that vodka protected people from radiation that was most interesting. It is, of course, a false claim, even if it does give hope to alcoholics! The dose prescription for the "medicine" was jokingly described in the following terms:

> Take a bottle of red wine and add 7 drops to a glass of vodka and then drink. Repeat this prescription until the bottle of wine is empty. You are then cured of radiation sickness, but will die of cirrhosis of the liver.

Besides the visit to the power station, I also visited Chernobyl town (where I was warmly welcomed by Mr Alexander Kovalenko* at the building which houses the Government Commission on the Liquidation of the Consequences of the Accident, headed by the Deputy Premier Mr Boris Scherbina), and the ghost town of Pripyat, which had 20 centimetres of topsoil removed and where, in windows of some flats, washing is still left hanging—a reminder of the necessarily hurried departure of the 45,000 population 19 months ago. I also learned that evacuation could not be undertaken by rail because of the high level of contamination of the station platforms.

Finally, Zelony Mys was visited, the new town built June to September 1986 and which can house 6000 power plant workers. It was only when we reached here at 1700 hours that we stopped to eat! My hosts had been so co-operative with my requests for site visits, photographs and information that there had been no time for other than my fact-finding exercise.

I returned from Kiev via Moscow on my way back to London and ended a memorable 6 days in the Soviet Union by having discussions with Mr Yuri Kanin†, the Moscow Novosti Press Agency's expert on Chernobyl; by meeting again Nikolai and Ludmilla Pakhomov‡; and by seeing the ballet *Giselle* at the Kremlin Palace, and the opera *Ivan Susanin* and the ballet *Swan Lake* at the Bolshoi.

Additional Statistics

- The evacuation column which set out from Kiev for the 130 kilometre journey to Pripyat after midnight on 27 April 1986 consisted of 1216 large buses and 300 trucks and stretched for 15 kilometres.
- 34,500 were evacuated in the buses from Kiev, whilst 9000 left in Pripyat city transport and in their own cars.
- Evacuation of the population from the 10-kilometre zone was completed by 2 May and from the 30-kilometre zone by 5 May. The total evacuees were some 116,000 persons and 86,000 head of cattle.
- State compensation cash benefit payments for evacuees were: 4000 roubles for single persons, 7000 roubles for a family of two and 1500 roubles for each additional family member.
- 1000 square kilometres of land around the nuclear power stations was to some extent contaminated.

* Head of Information and Foreign Relations Department, AI "Kombinat", Chernobyl.
† Managing Editor, Science and Technology Division, Novosti Press Agency, Moscow.
‡ Chief correspondent and picture librarian of the London bureau of TASS until September 1987.

- All Kiev streets, large and small, were continually washed throughout the summer of 1986 until bad weather prevented such washing. 300,000 tons of leaves were buried from Kiev in October and November 1986, beyond the city outskirts.
- The Soviet authorities are planning to publish a book in Russian in March 1988 entitled *Lessons of the Chernobyl Tragedy* (Political Publishers, Moscow), by E. Ignatenko, A. Kovalenko, S. Toitsky *et al.* It will be approximately 200 pages in length with some 60 illustrations and will provide answers to some 112 major questions concerning the accident and its consequences.
- The temperature inside the sarcophagus is now 82°C and the point was made by Mr Kovalenko that snow did not melt on its top whilst snow remained on the tops of units Nos. 1, 2 and 3—thus disproving any statement that the sarcophagus was "hot".

R. F. Mould

6 December 1987
Moscow

The author with Mr Mikhail Umanets, Director of the Chernobyl Power Station, 2 December 1987.

Control room of unit No. 1. The author is far left and Mr Umanets is third from the left.

Photographs by Mr Dmitri Chukseyev of Novosti Press Agency, Moscow

Measuring the radiation dose rate outside the sarcophagus, with a radiation monitor provided by Mr Kovalenko. The reading was 4.0 millirem per hour.

The radiation dose rate measured at this site was 16.3 millirem per hour, but was due not to radiation from the sarcophagus itself, but from a mound of debris on the shoulder of the road beyond the wall surrounding the power station. Measurement in Pripyat at a site near that shown in Fig. 155 was 0.2 millirem per hour. The trees in Pripyat had all been removed, but it was stated that in a forested area at some distance from Pripyat the dose rate was 80 millirem per hour.

Photographs by Mr Dmitri Chukseyev of Novosti Press Agency, Moscow

Acknowledgements

WITHOUT the invaluable support of my wife Maureen and my children Timothy, Fiona and Jane, this book could not even have been attempted. This support included practical help in identifying and videotaping TV news items and documentaries, searching newspapers for relevant material, transcribing tapes from the 25–29 August 1986 IAEA/Vienna meeting, French translations and constructive criticism.

Also, the final manuscript of this book would not have been completed without the assistance of many other people. In particular, I would like to thank the following who contributed to varying aspects in its production, such as provision of papers, reports, references, illustrations and helpful comments:

Mr U. Altemark, Prof. A. Baranov, Dr H. Bergmann, Prof. R. J. Berry, Mr A. Bose, Mr R. J. Chase, Mr S. Chasnikov, Prof. D. Chassagne, Mrs Angela Christie, Mr D. Chukseyev, Prof. L. Cionini, Dr R. H. Clarke, Mr A. Collings, Mr T. Copestake, Mr F. Cottam, Mr V. I. Dergun, Mrs Flora Dermentzoglov, Prof. Andrée Dutreix, Mr J. Dunster, Prof. S. Eckhardt, Lord Ennals, Dr A. E. J. Eggleton, Miss Janet Fookes MP, Miss Frances Fry, Mr T. Garrett, Mr R. Gibson, Dr J. H. Gittus, Prof. Angelina Guskova, Dr I. W. F. Hanham, Dr J. Hopewell, Mr R. Hurry, Prof. M. H. Husein, Prof. L. A. Ilyin, Mr M. Jones, Dr W. Jasinski, Mr Y. Kanin, Mr A. P. Kovalenko, Mr G. Lean, Mrs Alla Levchenko, Mr Y. Levchenko, Mr B. McSweeney, Lord Marshall of Goring, Miss Diana Makgill, Mr H-F. Meyer, Mr L. Meyer, Miss Rosemary Nicholson, Dr M. Nofal, Mr E. Oksyukevich, Mr V. Orlik, Mrs Lyudmila Pakhomova, Mr N. Pakhomov, Dr P. Paras, Mr E. Protsenko, Dr N. T. Racoveanu, Dr I. Riaboukhine, Dr H. Roedler, Mr G. Shabannikov, Mr Y. D. Shevelev, Mr J. Stenning, Prof. H. Svensson, Dr Y. Skoropad, Miss Gillian Smithies, Dr G. N. Souchkevitch, Dr L. Touratier, Dr N. G. Trott, Mr M. P. Umanets, Dr G. van Herk, Mr E. van't Hooft, Dr P. J. Waight, Dr W. S. Watson, Mr G. Webb.

The organisations which have also been most helpful include:

TASS in London
Novosti Press Agency in London
Novosti Press Agency in Moscow
Information & Foreign Relations Department AI
"KOMBINAT" in Chernobyl

Associated Press in London
World Health Organisation in Geneva
International Atomic Energy Agency in Vienna
British Embassy in Moscow
Soviet Embassy in London
National Radiological Protection Board in the United Kingdom
Central Electricity Generating Board in the United Kingdom
Kerntechnïsche Hilfsdienst GmbH, Eggenstein-Leopoldshafen, Federal Republic of Germany
Intourist in London
Foreign & Commonwealth Office in London
Department of Energy in London
Soviet Television
The British Broadcasting Corporation

I am particularly grateful to the following for facilitating my Soviet visa and for ensuring my visit to the Chernobyl power station on 2 December 1987 was a success: Mr Victor Orlik, Mr Dmitri Chukseyev, Mr Alexander Kovalenko, and the Director of the power station, Mr Mikhail Umanets. My publishing deadline (11 November 1987) had passed by the time of this visit, but I am grateful to Mr Barry Martell for still allowing the inclusion of an afterword, some additional figures, and a few footnotes in the chapters.

Finally, I would like to thank my Pergamon editorial, marketing and production colleagues, Myles Archibald, Glenda Kershaw, Brigitte Kaldeich, Clare Grist and Barry Martell, for their enthusiastic and helpful encouragement, and for allowing the author some elastic deadlines!

1

The Chernobyl Nuclear Power Plant

"Russia is communism and electrification."
(V. I. Lenin, 1870–1924)

THROUGHOUT THE world there are several different designs of nuclear reactor to power turbines to generate electricity for a national grid. The reactors at Chernobyl, of which there were four by April 1986, with two more planned for the future, are known as RBMK1000 and are only to be found in the Soviet Union. They have been developed from their first nuclear reactor power plant which was planned in the 1950s and commissioned at Obninsk, the Russian equivalent to the United Kingdom's Harwell, in June 1954. The old style reactor produced 5 megawatts (MW) of electrical power, whereas the RBMK1000 produces 1000 MW. The first two of the RBMK1000 power plants were built at Leningrad in 1973 and 1975, and from 1973 to 1982 ten were put into operation in the Soviet Union. In 1986 there were fourteen in service and eight under construction.

An RBMK1000 power plant is designed to have twin reactors, like semi-detached houses, with the two independent reactor systems having a number of interchangeable auxiliary systems. At Chernobyl the accident occurred in unit No. 4 and its twin, No. 3, was for some time also at risk due to the spread of burning debris. Units Nos. 1 and 2 had the outsides of their buildings contaminated. *Izvestia* on 3 October reported that unit No. 1 was restarted on 26 September and on 1 October "electricity was churned out again". However, in late November 1986 the status was that unit No. 1 was again shut down and is now "under revision", whereas Nos. 2 and 3 are "being prepared". This was from an ex-member of staff at Obninsk who is now on secondment to an international agency. Units Nos. 5 and 6 at the time of the accident[1]* were only a large hole in the ground and a series of architectural drawings. This was fortunate since the site of No. 5 provided an immediately available storage pit for contaminated material, solving a problem in the short-term but creating one for the long-term.

---

* Superscript numbers refer to notes added at proof stage (pp. 233–239).

The accident has seriously affected the electrical power production capability of the Soviet Union, not only by the permanent loss of unit No. 4 and possibly also of unit No. 3, but by the now pressing need to carry out safety modifications on all existing RBMK1000 reactors and to further evaluate Soviet plans to extend the RBMK1000 design to an RBMK1500 capable of producing 1500 MW of electrical power. This stark reality is emphasised by the statistics in an *IAEA Bulletin* article in 1983, entitled "Nuclear power in the Soviet Union". It states that "in 1980, the RBMK1000 reactors produced 64.5% of all electrical power produced by Soviet nuclear power plants, which represents 47 billion kilowatt-hours of a total of 73 billion kilowatt-hours". In the same article it was also said that the average operational load factor for the RBMK1000 plants at Leningrad, Kursk and Chernobyl was 75%. The Soviet Union can ill afford even the temporary loss of these power plants.

The RBMK1000 has been variously called a "boiling-water, graphite–uranium high-power reactor" (*IAEA Bulletin*, 1983) and a "thermal neutron channel-type (pressure tube) reactor" (USSR Delegation, 25–29 August 1986). The four major design features of RBMK1000 reactors which determine all the main characteristics of the reactor and the nuclear power plant are:

Vertical channels containing the fuel and coolant, enabling local refuelling while the reactor is in operation.

Fuel in the form of bundles of cylindrical fuel elements made of uranium dioxide in zirconium tube-type cladding.

A graphite moderator between the channels.

A boiling light-water coolant in the multiple forced circulation circuit, with direct steam feed to the turbine.

Prior to the accident, the structure of unit No. 4 at Chernobyl rose to a maximum height of 71.3 metres directly above the reactor core, the reactor hall and the massive robot crane used for the refuelling process. The maximum width of unit No. 4 was 191.3 metres. After the final reinforced concrete entombment of the reactor, the structure will be almost as monumental, still 60 metres high, although the tall refuelling crane now lies as useless twisted and broken metal at the bottom of the shattered reactor. The RBMK1000 reactor core is 11.8 metres in diameter by 7 metres high, and the fuel channels are located in 1661 cells in a square lattice. Each channel contains two uranium fuel assemblies and each fuel assembly consists of a cluster of eighteen fuel elements of 2% enriched uranium dioxide pellets in a zirconium alloy cladding. The pressure tubes are also made of a zirconium alloy.

Radioactive zirconium-95, with a half-life of 64 days, was one of the radioactive isotopes identified beyond the borders of the Soviet Union in the early days following the accident. In the United Kingdom, for example, two American lady tourists arrived in Glasgow on 1 July, having earlier decided to avoid London because of concern in the United States that Colonel Ghadaffi of Libya was about to bomb much of that city. They travelled instead to visit Kiev from 29 April to 2 May. Zirconium-95 was found in very small quantities on a pair of their shoes—which they hurriedly discarded! However, an interesting comparison was able to be made between the relative amounts of various radioactive isotopes on the "Kiev shoe" and on a "Warsaw T-shirt" which had been worn outdoors on 2 May and 5 May by a British tourist. It showed that the major deposition of particles of zirconium-95 occurred in Kiev in the Soviet Union, rather than farther afield. This was probably a result of zirconium being present in bulk in the reactor as a structural cladding constituent of the fuel rods.

In addition to the fuel channels already mentioned, there are also 221 channels for the reactor control and protection system. The graphite moderator stack housing fuel and control rods is located in a leak-tight cavity which is called the reactor space and this is filled with a helium–nitrogen mixture to prevent oxidation of the graphite. The control and protection system is obviously of vital importance and is based on the movement of the 211 boron carbide aluminium alloy-clad neutron absorber rods. These are in specially separated channels cooled by water from an autonomous circuit. If there are fewer than fifteeen control rods inserted in the reactor, then the written safety rules require the operators to trip (the term for shutdown) the reactor. It has been admitted (USSR Delegation, 25–29 August 1986) that this is a design fault in the RBMK1000 reactor. In practice at Chernobyl, the operators failed to trip the reactor in the emergency. An automatic trip mechanism would have prevented this from happening.

In the RBMK1000 the uranium fuel in the channels within the graphite moderator is cooled by boiling water, and this process produces the steam necessary to drive the turbines and generate electricity. There are four primary circulating pumps to pump water into the bottom of the channels and it boils as it rises to the channel tops where a mixture of hot water and steam emerges. This passes through pipes to steam separators, sometimes called steam drums, where the steam collects above the water and is led by more pipes to the turbines. The water is drawn off and pumped back through the channels to be boiled again. The steam from the turbines is condensed and also pumped through the channels to complete the cycle of operations.

2

Kiev

"How fortunate are those who farm in the Ukraine, living among their kingdoms of wheat."

(Honoré de Balzac, 1799–1850)

THOSE 135,000 people from the towns of Pripyat and Chernobyl and from the surrounding areas in the 30-kilometre evacuation zone around the nuclear power plant would not have echoed Balzac's opinion, but it is perhaps not realised that the toll of the accident could have been much worse, and that in this sense Kiev in particular was fortunate. For instance, the radioactive cloud did not initially pass over Kiev, as the prevailing wind was in a different direction, towards north-east Poland and Warsaw; nor did it rain for several days and consequently radioactivity was not deposited in rainwater but largely remained in the atmosphere; and, most importantly, the vast River Dnieper, which flows through the valley of the same name where 3.8 million people live, was not contaminated, as at one stage was considered likely.

Kiev, the capital of the Ukraine and a city of great historical importance, has a population of 2.5 million, which makes it the third largest city in the Soviet Union. It is the ancient holy city of Russia and in the first century was the capital of Kievan Rus when Moscow was little more than a village. It was also in Kiev in the first century that orthodox Christianity was introduced into Russia, the famous and still standing Cathedral of St Sophia was built, and the cyrillic alphabet was devised. Some aspects of modern-day Kiev can be seen from the illustrations, and the logistical and human problems which would have been encountered if any large-scale evacuation of Kiev had been necessary defy the imagination. The evacuation and rehousing of the 135,000 people from the 30-kilometre zone was a most impressive achievement by the Soviet authorities in so short a time, but if Kiev had also been involved, the chaos this would have caused would have been incredible.

One of the early fears for Kiev was the possible contamination of the water table below unit No. 4, as a result of a sort of Hollywood-style China Syndrome with the meltdown of the reactor core, followed by it burrowing its way through the foundations of the reactor building and through into the earth. This did not in fact happen, in spite of various media headlines such as the following.

Newsweek magazine, 12 May 1986

Under the headline "The Chernobyl Meltdown" was written: "The very word conjures up the horror, a reactor fire hot enough to turn metal to ooze, graphite to glowing charcoal [this actually happened to one-quarter of the graphite], and otherwise controlled elements to free floating isotopes of death." Another headline said "The 20th century plague."

The Guardian newspaper in the United Kingdom, 1 May 1986

The main headline was "Russians deny second meltdown" and a sub-heading was "US estimates up to 3000 victims from satellite[2] information."

The Sunday Times newspaper in the United Kingdom, 11 May 1986

In the "News in Focus" section of the paper was the headline "Cloud over Kiev", and a large black and white photograph with the caption "It was 3 pm on a sunny day when a tourist took this picture of the nuclear cloud, a cloud whose effects fill the residents of Kiev with fear." The photograph below this showed what looked like five rather bored students on benches against a wall. The skyline leaves little doubt that the photograph is of Kiev, but one must seriously consider whether a black cloud of soot and smoke from any massive inferno some 130 kilometres away (the distance between Chernobyl and Kiev) would appear like this. The tourist's picture also appeared in *Time*, 12 May 1986, but this time in colour, showing an impressive orange glow. I think that perhaps the truth is somewhat different, because I tracked down the colour transparency to the London agency representing the French photographic agency Sygma which held the exclusive. To my surprise there was no orange glow in the colour transparency, only rather insipid colours! The Sygma caption read as follows: "On Saturday 26 April 1986, 24 hours after the beginning of the accident in the nuclear plant at Chernobyl, and 48 hours before the catastrophe was made public by the Swedes, a black cloud darkened the sky over Kiev, 130 kilometres south of the plant. From a boat on the Dnieper, a German tourist photographed the highly impressive phenomenon. A document which allows the viewer to suppose that the Ukrainian capital has perhaps been struck by the consequences of the most dramatic accident in the history of civil nuclear energy."

One of the main worries immediately after the accident was contamination of Kiev's water supply, and V. I. Trefilkov, Vice-Chairman of the Ukrainian Academy of Sciences, gave an unscheduled report at the 25–29 August 1986 IAEA meeting, stating that in the first few days after the accident measures were taken to provide alternative water supplies, and that in Kiev 400 wells were bored to replace the water normally taken from the River Dnieper. Water purification techniques were also introduced, including the use of absorbents to reduce any radioactivity present by a factor of 100.

What might have happened to Kiev should serve as a lesson for the future on the siting of nuclear power plants within accidental reach of highly populated areas, and near rivers and reservoirs connected to drinking water supplies.

3

The Accident

"Read not to confuse, nor to believe and take for granted, nor to find talk and discourse but to weigh and consider."
(Francis Bacon, 1561–1626)

THE NUCLEAR power plant is situated in a flat landscape beside the 200–300 metre wide River Pripyat, a tributary of the Dnieper. The region is known as the Byelorussian–Ukrainian woodlands, because although actually in the Ukraine, it is very near the border with White Russia, whose capital Minsk, with its population of 1.3 million, is 320 kilometres from the power plant. The regional centre is the town of Chernobyl, which has a population of 12,500, and is situated 15 kilometres to the south-east of the plant. Nearer, only 3 kilometres from the plant, is the satellite town of Pripyat, with 45,000 inhabitants. It is a young town and its inhabitants are young, power plant engineers, building workers, chemists, river transport workers and their wives and children. Most, it can be assumed, regarded the presence of the four Chernobyl nuclear reactors as a routine fact of life. They were used to them and, after all, the power plant provided a livelihood for many. It is usually found that when sites for new civil nuclear power installations are discussed in public, the local opinion is vociferously against any site "on their own doorstep", although this can often be overcome. The now famous French ministerial remark given when asked about the siting of nuclear power plants in France—"You don't ask the frogs when you drain the marshes"—is not an ideal method! However, once a nuclear site is operational, providing a large proportion of jobs and income for the local population, as in the area around Sellafield (called Windscale at the time of the 1957 accident) in the United Kingdom, the locals became very protective of their new way of life. They react strongly to outside criticism. It was probably similar in Pripyat before 25 April 1986, when an experiment started at unit No. 4 prior to a scheduled shutdown for annual maintenance.

On the night of 25–26 April there were on the power plant site 176 duty operational staff and workers from various departments and maintenance services. In addition to this number, there were also 268 builders and assemblers working on the night shift of the constructions of units Nos. 5 and 6. This was the third construction stage, as units Nos. 1 and 2 were built in the first stage, 1970–7, and the second stage consisting of units Nos. 3 and 4 was completed in December 1983.

The fatal accident sequence was initiated by a decision of the plant's management and specialists to make an overnight experiment to test the ability of the turbine generator to power certain of the cooling pumps whilst the generator was freewheeling to a standstill after its steam supply had been cut off. The purpose of the experiment was to see if the power requirement of unit No. 4 could be sustained for a short time during a power failure. It has been admitted (USSR Delegation, 25–29 August 1986) that these tests were not properly planned, had not received the required approval, and that the written rules on safety measures said merely "that all switching operations carried out during the experiment were to have the permission of the plant shift foreman, that in the event of an emergency the staff were to act in accordance with plant instructions and that before the experiments were started the officer in charge [an electrical engineer, incidentally, who was not a specialist in reactor plants] would advise the security officer on duty accordingly". Also, that "apart from the fact that the programme made essentially no provision for additional safety measures, it called for shutting off the reactor's emergency core cooling system. This meant that during the whole test period [about 4 hours] the safety of the reactor would be substantially reduced." In addition, since "the question of safety in these experiments had not received the necessary attention, the staff involved were not adequately prepared for the tests and were not aware of the possible dangers". The nuclear power plant staff conducting the experiment, incredibly as it might seem, knowingly departed from the experimental programme which was already of a poor quality. This created the conditions for the emergency situation which finally led to the accident which no one believed could ever happen.

The following is a chronological account of what happened between 0100 hours on 25 April 1986, when power reduction for annual maintenance was started, and 0124 hours on 26 April 1986, when two explosions occurred about 3–4 seconds apart, throwing burning lumps of material and radioactivity into the atmosphere. That brief span of just over 24 hours will rank in nuclear history alongside the discovery of radioactivity by Henri Becquerel in Paris on 1 March 1896 some 90 years earlier, the Curies' discovery of radium in 1898, and the events at Hiroshima and Nagasaki in 1945; in that world perception, both lay and scientific, of nuclear possibilities has been ineradicably altered.

25 April

0100

Start-up of reactor power reduction in preparation for the experiments and the planned shutdown of unit No. 4.

1305

Reactor power reduced to 1600 MW of thermal power, which was 50% of the maximum thermal power of the reactor.* Unit No. 4 has two turbine generators, Nos. 7 and 8, and turbine generator No. 7 was tripped (terminology for shutdown)

* The specification of the plant includes the thermal power of the reactor and the electrical power which can be generated; both are measured in megawatts (MW). The maximum thermal power of an RBMK1000 reactor is 3200 MW and the maximum electrical power is 1000 MW.

from the electricity grid and all its working load, including four of the main circulating pumps, transferred to turbine generator No. 8.

1400

As part of the experimental programme, the reactor's emergency core cooling system was disconnected. However, at this point in time the experiment was subjected to an unplanned delay because of a request by the electricity grid controller in Kiev to remain supplying the grid till 2310 hours. This was agreed by the Chernobyl staff, but [**major fault no. 1**] the reactor's emergency core cooling system was not switched back on and this represented a violation of written operating rules for just over 9 hours.

2310

The reduction of the reactor's thermal power was resumed since, in accordance with experimental procedure, the test was to be performed at between 700 MW and 1000 MW thermal power.

26 April

0028

On going to lower power, the set of control rods used to control reactor power at high powers, and called local automatic control rods (LACs), were switched out and a set of control rods called the automatic control rods (ACs) were switched in. However, [**major fault no. 2**] the operators had failed to reset the set point for the ACs and because of this they were unable to prevent the reactor's thermal power falling to only 30 MW, a power level far below the 700–1000 MW intended for the experiment.

0100

The operators succeeded in stabilising the reactor at 200 MW thermal power, although this was made difficult due to xenon poisoning of the reactor. The 200 MW level was only achieved by removing control rods from the core of the nuclear reactor. Nevertheless, 200 MW was still well below the required power level and the experiment should not have proceeded—but it did.

0103 AND 0107

The two standby main circulating pumps were switched respectively into the left and right loops of the coolant circuit. Eight main pumps were now working and this procedure was adopted so that when, at the end of the experiment in which four pumps were linked to turbine generator No. 8, four pumps would also remain to provide reliable cooling of the reactor core. However, due to the low power of 200 MW and the very high (115–120% of normal) coolant flow rate through the core due to all eight pumps functioning, some pumps were operating beyond their permitted regimes [**major fault no. 3**]. The effect was to cause a reduction in steam formation and a fall in pressure in the steam drums.

0119

The operators tried to increase the pressure and water level by using the feedwater pumps. The reactor should have tripped because of the low water level in the steam drums, but [**major fault no. 4**] they had overridden the trip signals and kept the reactor running. The water in the cooling circuit was now nearly at boiling point.

0119.30

The water level required in the steam drum is reached, but the operator continues to feed water to the drum. The cold water passes into the reactor core and the steam generation falls further, leading to a small steam pressure decrease. To compensate for this, all twelve automatic control rods (ACs) are fully withdrawn from the core. In order to maintain 200 MW thermal power, the operators also withdrew from the core some manual control rods.

0119.58

A turbine generator bypass valve was closed to slow down the rate of decrease of steam pressure. Steam is not dumped into the condenser. Steam pressure continues to fall.

0121.50

The operator reduces the feed water flow rate to stop a further rise in the water level. This results in an increase in the temperature of the water passing to the reactor.

0122.10

Automatic control rods (ACs) start to lower into the core to compensate for an increase in steam quality.

0122.30

The operator looks at a printout of the parameters of the reactor system. These are such that the operator is required in the written rules to immediately shut down the reactor, since there is no automatic shutdown linked to this forbidden situation. The operator continues with the experiment [**major fault no. 5 and a most serious one**]. Computer modelling has shown that the number of control rods in the reactor core were now only six, seven or eight, which represents less than half the design safety minimum of fifteen, and less than one-quarter the minimum number of thirty control rods given in the operator's instruction manual.

0123.04

The experiment is started with the reactor power at 200 MW, and the main steam line valves to the turbine generator No. 8 were closed. The automatic safety protection system which trips the reactor when both turbine generators are tripped was deliberately disengaged by the operators [**major fault no. 6 and the most critical fault**], although this instruction was not included in the experimental schedule. After all, the operation of the reactor was not required after the start of the experiment. What seemed to be going through the mind of the operator was that if the experiment at first failed, then a second attempt could be made if the reactor was still running.

It is difficult to avoid the conclusion that the major priority of the Chernobyl unit No. 4 operators was to ensure that they completed the experiment during the 1986 rundown to annual maintenance and did not have to wait another 12 months before the next planned maintenance in 1987. It is hard to imagine a situation where the pressure and stress exerted on experimentalists is such that they would ignore many vital safety procedures just so that their tests could be completed. Nevertheless, this is just what happened.

0123.05

The reactor power begins to rise slowly from 200 MW.

0123.10

The automatic control rods (ACs) are withdrawn.

0123.31

The main coolant flow and the feedwater flow are reduced, causing an increase in the temperature of the water entering the reactor and an increase in steam generation. The operators noted an increase in reactor power.

0123.40

A reactor power steep rise (sometimes termed a "prompt critical excursion") was experienced, and the unit shift foreman ordered a full emergency shutdown. Unfortunately, the order came too late. Not all the automatically operated control rods reached their lower depth limits in the core and an operator unlatched them in order to allow them to fall to their positional limits under gravity. However, since the rods were nearly withdrawn, a delay of up to 20 seconds would have had to occur before the reactor power could have been reduced. This would have been 0124.00.

0123.43

Emergency alarms operate, but unfortunately the emergency protection is not sufficient to stop reactor runaway. The sharp growth of the fuel temperature produces a heat transfer crisis. Reactor power reaches 530 MW in 3 seconds and continues to increase exponentially.

0123.46

Intensive generation of steam.

0123.47

Onset of fuel channel rupture.

0123.48

Thermal explosion. According to observers outside Chernobyl unit No. 4, two explosions occurred at about 0124, one after the other; burning debris and sparks shot into the air above the reactor, and some of this fell on to the roof of the machine room and started a fire (USSR Delegation, 25–29 August 1986).

SUMMARY ANALOGY OF THE ACCIDENT

> "Imagine personnel of a plane which is flying very high. Whilst flying they begin testing the plane, opening the doors of the plane, shutting off various systems. . . . The facts (i.e. of the Chernobyl accident) show that even such a situation should have been foreseen by the designers."
>
> (Valeri Legasov, USSR Delegation Leader, 25–29 August 1986, IAEA, Vienna)

Many, but not all, newspaper, radio and television reporters like to embroider disaster stories with as many gory details as possible, in some cases true or untrue, and the initial response to the Chernobyl accident was no exception.

Daily Mail newspaper in the United Kingdom, 30 April 1986

The main headline was "2000 dead in atom horror" and a sub-headline said "Reports in Russia danger zone tell of hospitals packed with radiation accident victims." The source of the 2000 dead statistic was a supposedly intercepted radio ham message from Kiev which was picked up in The Netherlands. The figure of 2000 dead was to be often repeated in the early days following the accident.

Daily Mirror newspaper in the United Kingdom, 30 April 1986

The main headline was "Please get me out Mummy" and the sub-headline was "Terror of trapped Britons as 2000 are feared dead in nuclear horror."

American and Italian television networks, May 1986

TV pictures, supposedly of the Chernobyl plant burning, turned out to be a fire at a cement factory in Trieste, Italy. A TASS report from Rome, 15 May 1986, gave some relevant facts. Italian police arrested Thomas Garino, the maker of a fraudulent videotape shown on NBC and ABC television networks in the United States and on Italian state television. Italian viewers and TV journalists recognised the smoke-filled building as that of a Trieste factory. The film was taken in Trieste's industrial zone during the factory fire and at a local hospital (this was the Ospedale di Gattinara and pictures were also published in the local newspaper *Il Meridiano*) where the injured were taken. Garino is a Frenchman who posed as an Eastern European citizen back from a tourist trip to the Soviet Union, who had acquired the videotape sequence from a worker at the Chernobyl nuclear power plant. Garino collected a fee of US$20,000 and an ABC television newscaster later told his audience that "It is one mistake we will try not to make again."

Most cartoons about the Chernobyl accident which appeared in newspapers and magazines were either anti-Soviet, anti-American, anti-civil nuclear power or anti-nuclear war. However, a few only emphasised what are considered to be particular national characteristics. One such example was published in an Athens daily newspaper on 8 May 1986. Greeks, one is led to believe, are supposed to be pessimistic and tragedarian. Thus the Chernobyl accident was regarded as yet another national catastrophe and the panic buying of tins of food in the Athens shops reached enormous proportions with shoppers acting like unruly football crowds. In this cartoon the car boot is filled to overflowing with tins of evaporated milk and the passenger says: "What do I have to pay?" The taxi driver replies: "Three evaporated milks." The heading beneath the cartoon states that unbelievable quantities of food which could be stored were bought, and that all Athenians went to market!

Η ΚΑΘΗΜΕΡΙΝΗ

ΑΘΗΝΑ ΠΕΜΠΤΗ 8 ΜΑΪΟΥ 1986

(Σκίτσο τοῦ Κ.—)

ΑΠΙΣΤΕΥΤΕΣ ΠΟΣΟΤΗΤΕΣ ΣΥΝΤΗΡΗΜΕΝΩΝ ΠΡΟΪΟΝΤΩΝ ΚΑΤΑΝΑΛΩΘΗΚΑΝ

Μαζική ἔφοδος τῶν Ἀθηναίων χθὲς στὰ "σοῦπερ - μάρκετ"

ΑΓΟΡΑΖΑΝ Ο,ΤΙ ΕΥΡΙΣΚΑΝ ΜΠΡΟΣΤΑ ΤΟΥΣ, «ΜΗ ΕΠΙΚΙΝΔΥΝΟ»...

[*See caption opposite*]

New York Post, May 1986

This carried the most outlandish headline of them all, "15,000 dead in mass grave."

Detroit Medical News, 12 May 1986

The editorial stated: "So the Russians have started to self-destruct. That is the good news. The bad news is that they are exporting the fallout across the globe to us."

New Scientist, 4 September 1986

The headline was "How Chernobyl almost emulated Hiroshima" and this was introduced by "The Chernobyl reactor came close to exploding like an atomic bomb, engineers in Vienna were shocked and surprised to learn last week." In fact, at the 25–29 August 1986 IAEA meeting in Vienna, to which the *New Scientist* refers, the press briefing made it crystal clear that the nuclear power plant did **not** explode like an atomic bomb and that it was a steam explosion involving nuclear materials and not a nuclear explosion. Many journalists chose to ignore this information. However, it should be remembered that not all reporters and editors include unchecked sensational stories.

The Economist, 30 August–5 September 1986

This contained, under the heading "Tales from the Ukraine: the case for nuclear power has been dented but not destroyed", the following: "Remember that although reactors are designed by brilliant engineers, they are run by humble technicians who are used to cutting corners and who seldom understand how reactors work. At Three Mile Island technicians also overrode safety systems that baffled them." *The Economist* also added that "human error, though, is not the whole story".

The Washington Post, 29 April 1986

Headlines on the front page were "Soviet nuclear accident sends radioactive cloud over Europe", "TASS says mishap near Kiev caused unspecified casualties" and "Partial core meltdown suspected".

USA Today, 1 May 1986

Headline to a summary diagram stated "Soviet breadbasket threatened" and was followed by "The Ukraine comprises less than 3% of the Soviet Union, but produces 23% of its food. The Chernobyl disaster endangers the safety of food from this prime farming area." This was elaborated by the statistics "Sugar beet: 58% from Ukraine of the total Soviet production of 85.3 million tons"; "Grain: 23% from Ukraine of the total Soviet production of 158 million tons"; "Milk: 23% from Ukraine of the total Soviet production of 97.9 million tons"; "Potatoes: 23% from Ukraine of the total Soviet production of 85.5 million tons".

The Times, London, 30 April 1986

Front page headlines stated: "Fears of high death toll in atom disaster", "Poland sets up crisis team", "Russians end news blackout", "Failure to alert criticised" and "Future power stations could go back to coal".

The Times, London, 1 May 1986

Front page headline of "Chernobyl's second reactor threatened" with a sub-headline of "US Intelligence reports* disputed by Moscow".

Daily Telegraph newspaper in the United Kingdom, 29 April 1986

Front page headline of "Soviet atom leak alarm" with minor headings of "Fall-out 1000 miles across Europe", "World's worst reactor accident", "Britain not at risk" and "Echoes of U.S. disaster".

What had actually happened, triggered by the incomprehensive behaviour of the operators and design faults in the RBMK1000, was not a nuclear bomb explosion but a *steam* explosion caused by dispersed fuel in contact with water. Energy was deposited into the fuel at a high rate where it could not escape, and this probably resulted in melting of the fuel or of fuel fragmentation and dispersion. Ejected fuel hitting the pressure tubes would cause breaks in these tubes and steam would be released into the graphite moderator space. Water could then rush into the fuel channels and cause a fuel–coolant interaction, producing the first explosion. The second explosion could have been from hot hydrogen (produced by the oxidation of zirconium) and carbon monoxide (produced by the interaction of water with hot graphite) mixing with air when the top of the reactor vault was blown off.

Some of the first photographs released after the accident are shown in the section of photographs contained in this book. It is presumed that at least some of these were taken from a military helicopter. On 25 August 1986, the first day of the IAEA Post-Accident Review meeting in Vienna, near the start of the first Soviet delegation presentation, an overview of the accident was given by Academician Valeri Legasov, the delegation leader. At this time, many of the 500 delegates were not exactly expecting a wealth of hitherto unknown information. A colour videotape showing was announced by Legasov. In parts it was one of the most dramatic I have ever seen at a scientific meeting. Apparently its production had only been completed just before the Soviet delegation left Moscow for Vienna. The quality of the video was perfectly adequate, but from time to time the frames wobbled, presumably because the helicopter pilot and film crew were rightly anxious to be finished and away from the Chernobyl plant.† However, it was when the video frames started to show aerial views (see Fig. 18) deep inside the stricken reactor that the audience of 500 delegates held their breath in unexpected shock. Nobody had realised that such pictures existed, or indeed could have been taken. The sequence which zoomed down into the remains of the shattered unit No. 4 reactor space commenced with a lot of dark sand-coloured wreckage of bricks and twisted and mangled metal structures, until in one corner of a frame there appeared a bright scarlet mass. This was the burning graphite of some of the remains of the reactor core. It was like looking down the mouth of a small volcano. The video panned to more wreckage and the only colour

* See January 1988 Update, note 2 (p. 233).

† I was told that one month after the accident when a TASS photographer overflew the site, his hat blew off in the wind, and on landing his head was shaved as a safety measure.

which appeared other than brown was the yellow of four, apparently virtually undamaged, circulation pumps.

I think it is true to say that this now famous videotape altered beyond recall the atmosphere of the Post-Accident Review Meeting, while cynics, including at least one who was born in Eastern Europe, may claim this was its intention and that the entire Soviet presentation was just one massive exercise in disinformation. It was generally recognised that the Soviet delegation was going to present a very full and frank report on the accident. This indeed occurred (with a few exceptions) and was accompanied by the distribution to all delegates of over 500 pages of documentation (see Appendix 1), and photocopies of all slide illustrations except those of Professor Guskova who showed medical pictures of the firemen victims from Chernobyl (see Chapter 4 and Figs. 59, 60 & 61). The videotape was by popular request given a second showing at the end of the meeting on 29 August 1986, and was then presented by Academician Legasov to Dr Hans Blix, Director General of the IAEA. However, there was a caveat that the IAEA must not distribute copies of the videotape, or individual freeze frame pictures, and these dramatic pictures were only seen by relatively few. However, some of these pictures later appeared in the documentary "The Warning", shown on Soviet television on 18 February 1987.

The Soviet response to the accident was necessary and impressive in terms of containment and alleviation of its consequences within the borders of the Soviet Union. Nevertheless, it leaves one wondering whether, if an accident of similar proportions (remembering that RBMK1000 reactors are only in the Soviet Union) had occurred in other countries, particularly those which are small in geographical terms, these other countries could have coped with the problems—not least with evacuating and rehousing 135,000 persons.

The chronology of the emergency response was as follows.

26 April

0123.48

Thermal explosion and outbreak of fire in over thirty places due to high-temperature nuclear reactor core fragments falling on to the roofs of buildings adjacent to the now destroyed reactor hall. Diesel fuel and hydrogen stores were also threatened and firefighting took precedence over radiation protection, since an even bigger disaster would have occurred if the fires had got out of control.

0130

The three staff on duty at the Chernobyl site medical centre were alerted and the Moscow emergency centre was notified using the code words Nuclear, Radio-activity, Fire. What followed was described by the USSR Delegation, 25–29 August 1986: "Because of damage to some oil pipes, electrical cable short circuits and the intense heat radiation from the reactor, foci of fire formed in the machine hall over the turbine generator No. 7, in the reactor hall and in the adjoining, partially destroyed, buildings. At 0130 the firemen on duty, with the subsection of the fire division responsible for the power station, set out from the towns of Pripyat and Chernobyl to the scene of the accident. In view of the immediate threat that the fire would spread along the top of the machine hall to the adjoining unit No. 3, and as it was rapidly increasing in strength, the first set of measures taken was directed towards putting out the fire in this critical area. It was therefore decided that the fires inside the buildings should be put out with fire extinguishers and the fire hydrants installed inside."

0145

Two specialised teams of medical staff (later there were to be other teams) left Pripyat and 115 beds were made available in regional hospitals.

0210

The main fires on the turbine generator machine hall roof were extinguished. The first twenty-nine casualties were admitted to hospital.

0230

Fires on the roof of the reactor sector were extinguished.

0300

Potassium iodate tablets were distributed to all Chernobyl power plant workers and to all patients.*

0500

All local fires extinguished, although the graphite fire in the reactor core was still continuing. Different measures had been taken for different areas where fire was involved: water for cable rooms, gas for control rooms, and foam where oil was present. Two videos were shown during 25–29 August 1986; one has already been described. The other was shown in the exhibition area outside the main conference halls. This one was, I believe, made from extracts from Soviet Television news reports (I was allowed to tape record the entire soundtrack) and it included the following text: "Firemen were the first to bear the brunt of the accident. The flames had fanned out towards a nearby reactor and could affect the plant's network of cable ducts. Twenty-eight people fought the fire, some of them are no longer with us. Here is an extract from the firemen's report prepared in hospital. Under the circumstances no one showed slackness; on the contrary, everyone displayed solidarity and discipline and an ability to take independent and sometimes risky decisions—the only right decision in that situation. Everyone realised what was in store for them. In a critical situation the blaze had been extinguished by 5 o'clock.

---

* This was 1½ hours post-accident and Academician Ilyin on 26 August 1986 compared this to the 6-hour post-accident response for iodine tablet distribution at Three Mile Island, Harrisburg, USA, in 1979. However, it was noticeable during 25–29 August that the Soviet delegation made very few references to either nuclear warfare or to Three Mile Island (they probably realised it would be counter-productive!) and the first mention of the Windscale 1957 accident in the United Kingdom was by the Director of the National Radiological Protection Board, Harwell, Mr John Dunster.

This lack of repetitive Soviet referral to Three Mile Island and Star Wars altered some of the initial attitudes of the delegates, which can be summarised in the rather scandalous anecdote which was circulating at the start of the Post-Accident Review meeting. "The Americans are not talking to the Russians because of Star Wars; the French are not talking to the Russians since because of their high dependence on civil nuclear power they do not want to admit the existence of the word Chernobyl; the Germans are not talking to the Russians because of their Green political party; the —— (not quoted for obvious reasons!) are not talking to the Russians because they can never make their minds up about anything; so it is left to the British to talk to the Russians."

The plant and surrounding areas are cordoned off. People are not allowed into the contaminated area."

While all the firemen had stuck to their posts in these fraught circumstances, there are reports of a few people during the evacuation who had their own interests at heart rather than co-operating in a team effort. One woman worker, for example, ran away from her post to her parents' home. There have also been reports, such as in *The Times*, 16 June 1986, which quoted from *Pravda*, of senior officials being sacked and disciplined. These included the sacking of the Chernobyl nucler power plant director,[3] V. Bryukhanov, and chief engineer, N. Fomin, because of "their irresponsibility and poor leadership". In addition, *The Times* stated that "three other leading officials at the plant were strongly criticised by name, including a deputy director who abandoned his post at the most difficult moment". It was noticeable during the 25–29 August 1986 meeting that the Soviet delegation did not include a single person from the staff of the Chernobyl nuclear power plant. Was this because all senior management and scientists/engineers had been sacked?

Later reports also detailed the sacking of certain Soviet ministers. *The Times*, 20 July 1986, reported that criminal proceedings had been initiated against an unspecified number of workers. The Politburo had also sacked four top officials connected with the accident. They were named as: Y. Kulov, Chairman of the State Atomic Power Inspectorate; G. A. Shasharin, a Deputy Minister of Power Engineering and Electrification; A. Meshkov, First Deputy Minister of Medium Machine Building; and V. S. Yemelyanov, Deputy Director of a research and design institute. The Minister of Power and Electrification,[4] A. Mayorets, received only a rebuke because he had held his post only for a short time.

26 April

0600

Confusion over messages sent to the Moscow emergency centre. Apparently reports after 0124, when the explosions took place, still indicated that the reactor was under control. By 0600 108 people were already hospitalised and one person had died from severe thermal burns. In the final analysis, only five received significant thermal burns. The burns which caused the threat to life were almost always beta-radiation burns, a fact not realised before the Chernobyl accident.

0640

Special emergency teams of physicists, dermatologists, radiologists and clinicians alerted in Moscow.

1100

The special emergency teams fly from Moscow to Kiev.

2000

In Moscow the Council of Ministers set up a government commission, incorporating leading scientists, specialists and officials, to deal with the accident and to determine its causes. The commission arrives to take charge at 2000.

2100

The emergency efforts were co-ordinated from a headquarters at the regional party committee building in Chernobyl. By 2100 (USSR Delegation, 25–29 August 1986) "an attempt has been made to reduce the temperature in the reactor vault and to prevent the graphite structure igniting, by using the emergency auxiliary feed pumps to supply water to the core space. This attempt proved ineffective. One of two decisions had to be taken immediately: (1) to contain the accident at source by covering the reactor shaft with heat-absorbent and filtering materials; or (2) to allow the combustion processes in the reactor shaft to come to an end of their own accord. Alternative (1) was chosen, since (2) carried within itself the danger that a significant area would suffer radioactive contamination, and the health of the inhabitants of major cities might be threatened."

Decisions were taken on the evacuation of Pripyat. The initial radioactive plume which had spewed out of the reactor building into the atmosphere had missed Pripyat, and now it was realised that original evacuation plans could have taken the evacuees into an area of higher contamination than Pripyat itself. The plans were re-evaluated.

27 April

0113

Chernobyl unit No. 1 was shut down.

0213

Unit No. 2, the last of the four plants still operating, was shut down. "Units Nos. 1, 2 and 3 and the power station equipment are checked by the staff on duty. Significant radioactive contamination of the equipment and buildings of units Nos. 1, 2 and 3 was caused by radioactive materials coming through the ventilation system, which continued to operate for some time after the accident. There was a significant degree of radiation in some parts of the machine hall, which was contaminated through the damaged roof of unit No. 3. The government commission ordered that decontamination and other work should be carried out on units Nos. 1, 2 and 3 with the aim of preparing them eventually for start-up and operation again" (USSR Delegation, 25–29 August 1986).

1400

Evacuation of the town of Pripyat was publicly announced and 40,000 people left the town in 2 hours 45 minutes.*

14 May

Soviet television speech by General Secretary M. S. Gorbachev. (The full text, given also to the Council of the World Health Organisation on 16 May, is reproduced in Appendix 2.) This was preceded in the Soviet Union by various reports in *Pravda* and *Izvestia*:

* The figure of 40,000 was quoted in the Soviet Television video, but the figure presented on 25–29 August was 49,000 for the evacuation of Pripyat.

Пролетарии всех стран, соединяйтесь!

Коммунистическая партия Советского Союза

ПРАВДА

Газета основана 5 мая 1912 года В. И. Лениным

Орган Центрального Комитета КПСС

№ 207 (24829) • Суббота, 26 июля 1986 года • Цена 4 коп.

7 May

Pictures of the helicopter pilots at Chernobyl and of medical checks and a report of 5000 evacuees sent to villages in the Borodyanskii region.

11 May

Pictures of a deserted building in Chernobyl and a bus with equipment used for "atmospheric purity control".

ИЗВЕСТИЯ

Газета выходит с марта 1917 года

СОВЕТОВ НАРОДНЫХ ДЕПУТАТОВ СССР

№ 164 (21606) ◆ Пятница, 13 июня 1986 года ◆ Цена 4 коп.

7 May

Brief mention of Chernobyl, but no photographs.

8 May

Longer article entitled "Events at Chernobyl atomic energy plant". No photographs.

9 May

Article entitled "The difficult hour". No photographs.

10 May

Article on the fire brigades entitled "Fulfilling his duty" and a photograph of Major Leonid Telyatnikov (see Chapter 4).

11 May

Article entitled "Changes come to the reactor". Photograph of a break in the work, showing a queue for drinks of the popular mineral water kvas.

13 May

Interviews and the TASS photograph of the damaged reactor with the arrow superimposed (see Fig. 19). This was the first photograph released by TASS.

"A peasant won't cross himself until he's hit by a thunderbolt."
[Quoted by Dmitry Kazutin, the *Moscow News* political observer, when writing about the Chernobyl accident. Kazutin suggests that this saying was probably coined even before the time of Tsar Peter the Great (1672–1725).]

Illustrations

FIG. 1. The Kreschatik, the main street of Kiev, in 1982, with its shops, cafes, cinemas, administrative buildings and a concert hall.

Courtesy of TASS

FIG. 2. The Kreschatik in 1943. This is not a nuclear landscape, but the aftermath of the German invasion of the Ukraine during World War II, which is known in the Soviet Union as the Great Patriotic War.

Courtesy of TASS

FIG. 3. Wedding couple in Kiev, May 1986.

Courtesy of TASS

FIG. 4. May Day 1986 celebrations in the village of Bobrovitso, Chernigov region in the Ukraine.

Courtesy of Associated Press

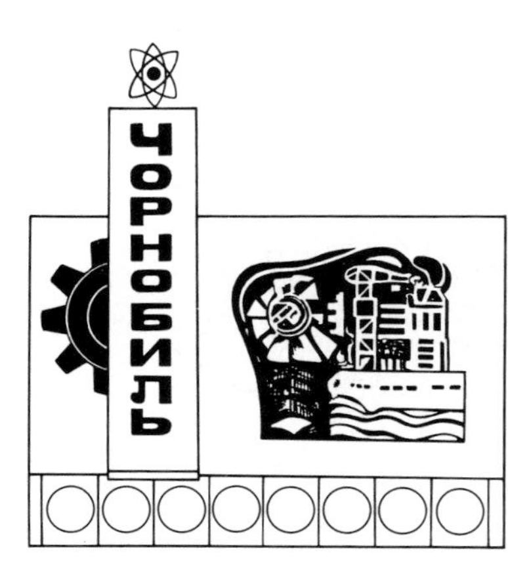

FIG. 5. The figure is a drawing of a distinctive multi-coloured roadside sign some kilometres from the Chernobyl atomic power plant, from a photograph in *Izvestia*, 19 May 1986, and still standing on 2 December 1987. After the accident, all the road shoulders in the 30-kilometre zone were asphalted and 2500 road signs were placed in position to denote the safest routes and to prohibit traffic in areas of high radioactive contamination.

FIG. 6. The Chernobyl power plant before the accident. This photograph was published in the February 1986 issue of *Soviet Life*.

Courtesy of Associated Press

FIG. 7. Simplified schematic diagram of a RBMK1000 nuclear reactor. The British Government's Energy Secretary, Mr Peter Walker, was informed during his 16–19 December 1986 visit to Chernobyl that, although the existing RBMK reactors would be modified and retained in use, no more RBMKs would be built in the USSR after completion of Chernobyl's units Nos. 5 and 6. Future reactors would be a pressurised water (PWR) design.

Refuelling machine
Control rods
Steam drums
Concrete shield
Steam drum
Turbine generator
Pressure tubes
Fuel elements
Condenser
Pumps
Feed pumps
Graphite moderator
Reactor core

FIG. 8. Photograph of all four units of the Chernobyl nuclear power plant as shown by the Soviet delegation at the 25–29 August 1986 post-accident review meeting at the IAEA, Vienna.

Courtesy of TASS

FIG. 9. A control panel at the Chernobyl plant.

Courtesy of The John Hillelson Agency Limited

FIG. 10. Photograph published in the February 1986 issue of *Soviet Life*. It shows workers at the Chernobyl nuclear power plant. The overprinted English text would not have appeared in the Russian language magazine and must have been added later. The cyrillic lettering over the door translates as "Strengthened peace by one's own work".

Courtesy of Popperphoto

FIG. 11. The shift chief in the reactor shop of Chernobyl unit No. 1 checks the radiation level at the rods' heads of the reactor, June 1986.

Courtesy of TASS

FIG. 12. The central hall of Chernobyl unit No. 1, June 1986.

Courtesy of TASS

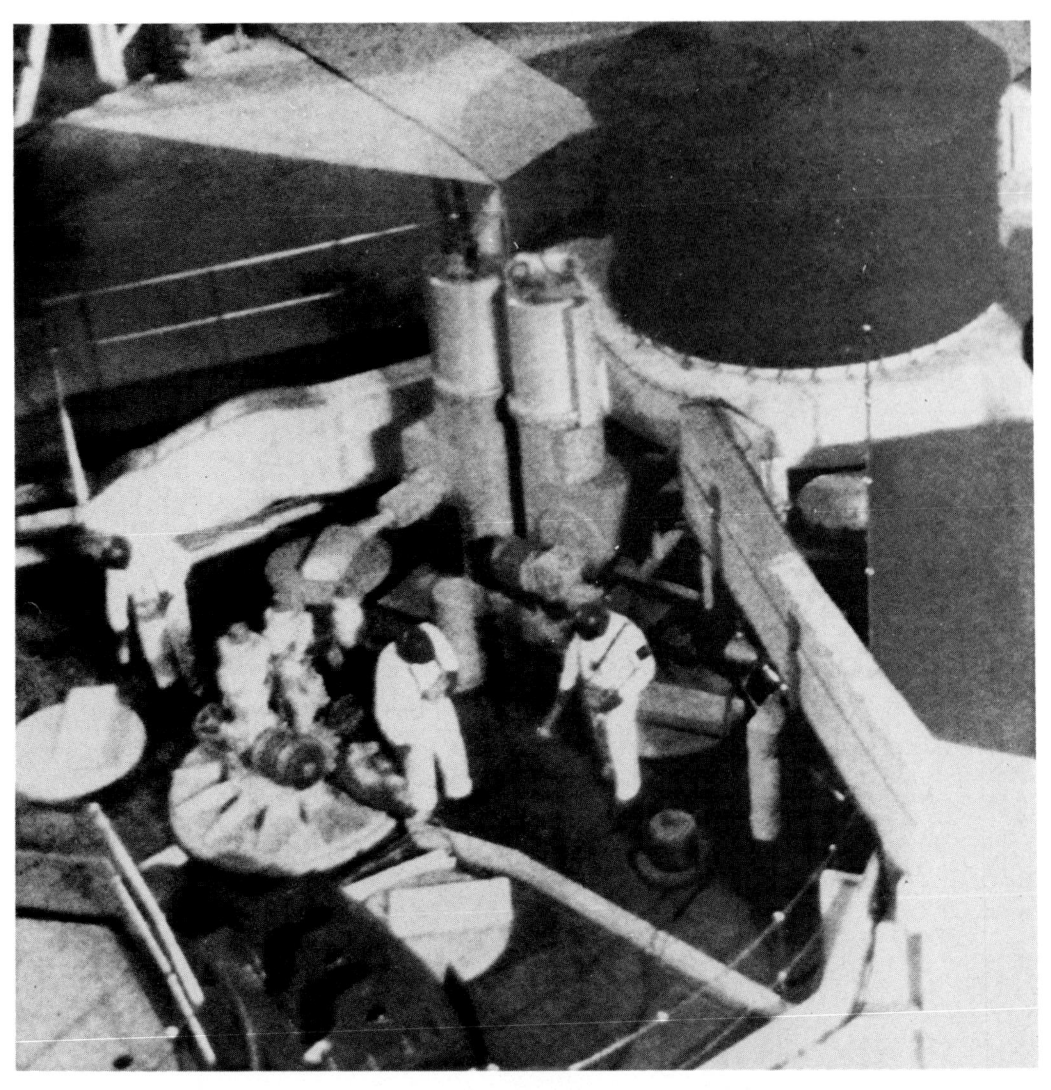

FIG. 13. Technicians in the reactor core cooling system complex of the Chernobyl nuclear power plant. This photograph was published in *Soviet Life*.

Courtesy of Associated Press

Fig. 14. The turbine house at Chernobyl, 1982.

Courtesy of TASS

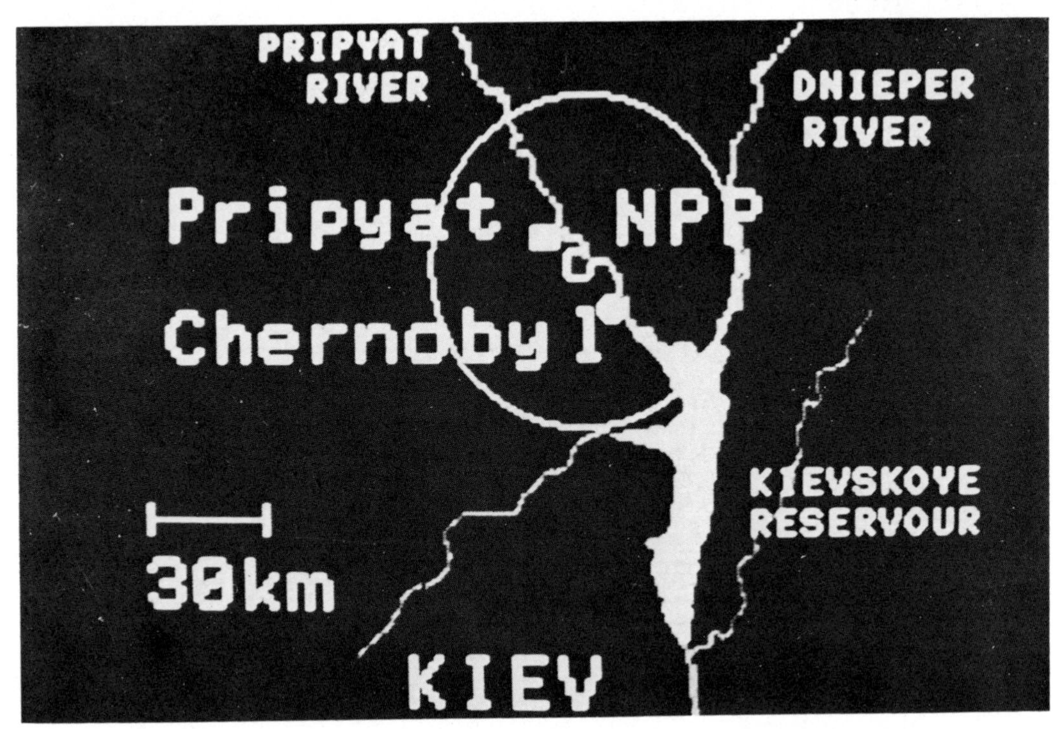

Fig. 15. Map of the area around the Chernobyl power plant as displayed by the Soviet delegation at the 25–29 August 1986 meeting of the IAEA in Vienna.

Courtesy of IAEA

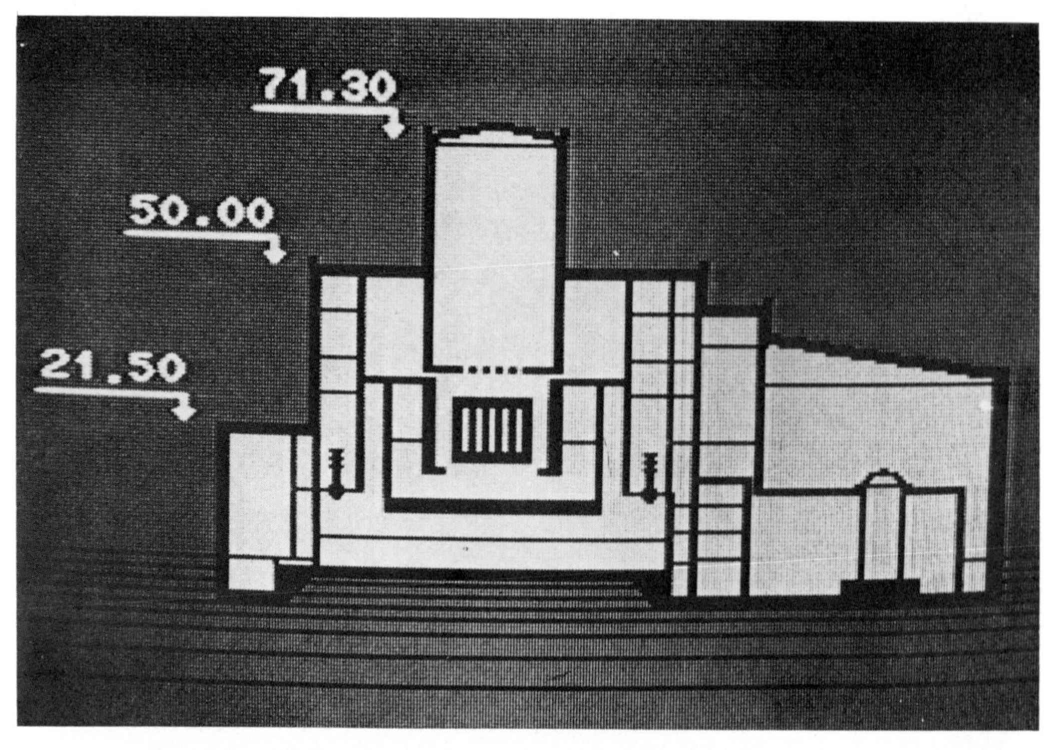

FIGS. 16 & 17. Cross-sectional view through unit No. 4 before and after the accident. The dotted lines show the outline of the planned entombment with a reinforced concrete sarcophagus.

Courtesy of IAEA

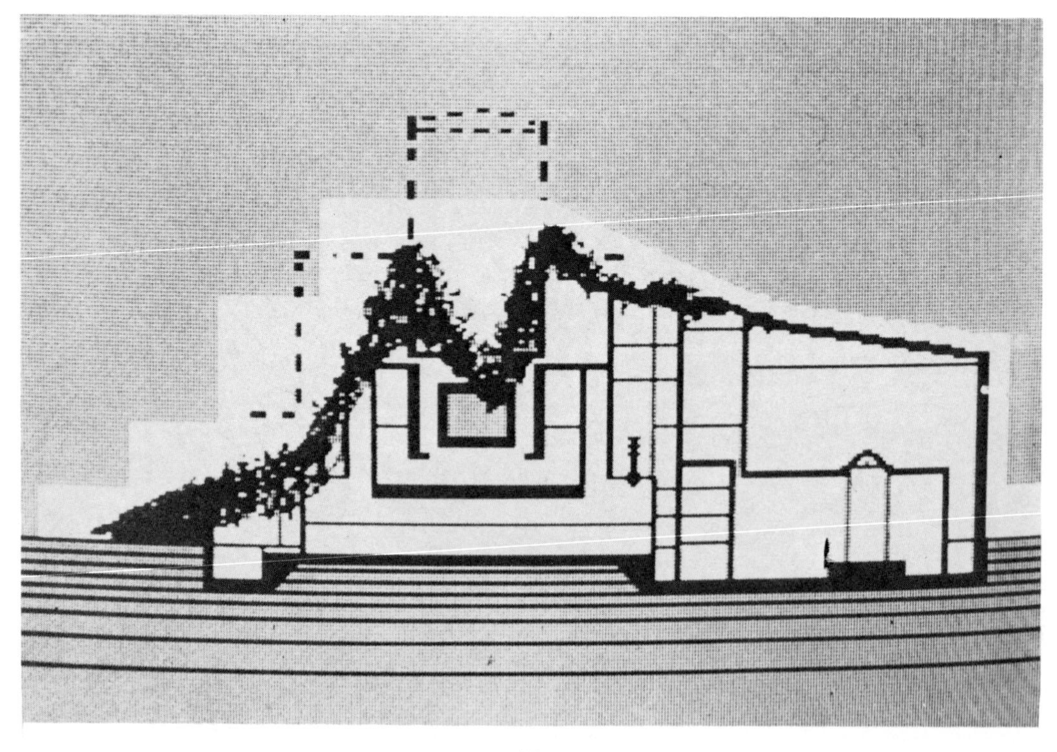

Fig. 18. View looking down the red and white chimney, with the smoke of the fire still showing.

Courtesy of Soviet Television from the film The Warning

Fig. 19. Aerial view of the crippled reactor, photographed on 9 May, and the first picture released by TASS to photographic agencies in Western countries.

Courtesy of TASS

FIG. 20. Remote control robots were used in the clean-up operations, but in some instances, such as on the warped roof of the reactor building, they were helpless. The brunt of this part of the clean-up had to be undertaken by men rather than robots.

Courtesy of Novosti

FIG. 21. A group of workers on their way to the destroyed roof of the reactor building, the highest radiation zone in which workers were to be allowed. The protective suits are constructed of lead and rubber.

Courtesy of Novosti

FIG. 22. Helicopter and bags of silicates/dolomite/lead ready for "bombing missions" to the damaged reactor.

Courtesy of Soviet Television from the film The Warning

FIGS. 23 & 24. *Left*: Helicopter pilots receiving instruction before flying to Chernobyl from their base in Chernigov, May 1986. *Right*: View through the helicopter windows of one of the loads scheduled for dumping on top of the stricken reactor, May 1986.

Courtesy of Pravda

FIG. 25. Helicopter treating the contaminated site with a deactivating solution which helps to neutralise the radioactive dust, May 1986.

Courtesy of TASS

FIG. 26. View of unit No. 4 through a helicopter doorway. The debris on the lower level roof can clearly be seen, September 1986.

Courtesy of TASS

FIG. 27. Major-General Antoshkin of the Air Force (see also Fig. 129) photographed with some of his helicopter pilots, December 1986.

Courtesy of Izvestia

FIG. 28. Commander of the helicopter MI-8, Captain Casimir Blin, 27 May 1986.

Courtesy of Izvestia

FIG. 29. Tunnelling under the damaged unit No. 4 to prepare the massive concrete slab containing heat exchangers which would ensure that the water table beneath the power plant would not be contaminated. Two miners can be seen manually pushing the truck.

Courtesy of Soviet Television from the film The Warning

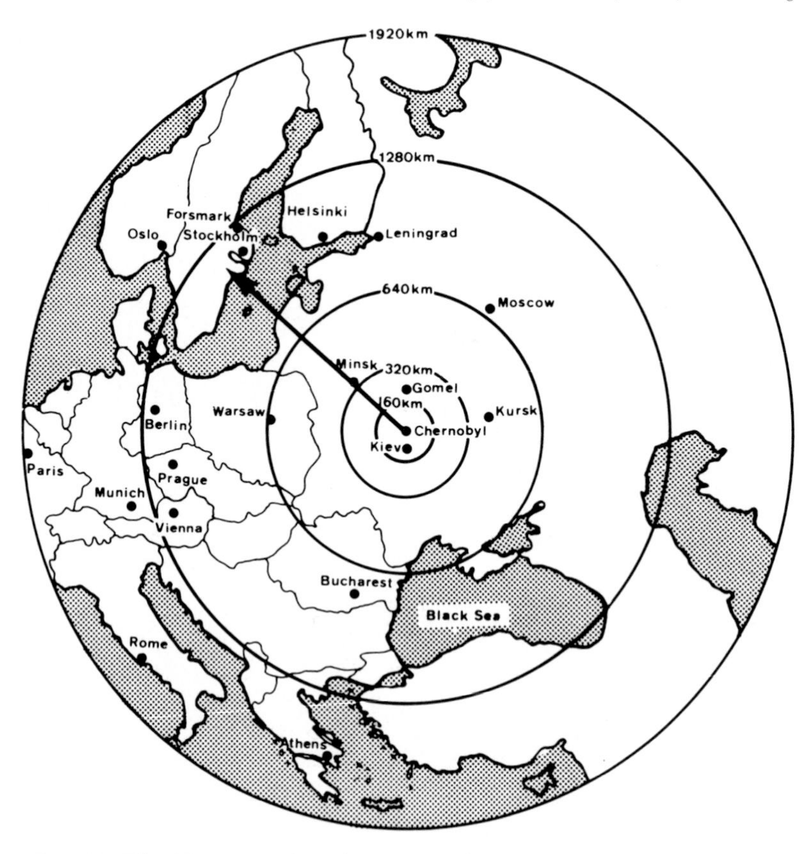

FIG. 30. The Ukraine and Byelorussia in the Soviet Union and surrounding countries. Fallout from Chernobyl was first detected at the Forsmark nuclear power station in Sweden. The arrow shows the initial direction of the radioactive plume. Appendix 3 details cancer incidence statistics for the Eastern European and Scandinavian countries shown in the map.

FIG. 31. The town of Pripyat, deserted after the evacuation, with the Chernobyl power plant in the background.

Courtesy of Soviet Television from the film The Warning

FIG. 32. Photograph of the Chernobyl nuclear power plant on 29 April 1986, taken from a Landsat 5 satellite. The lake in the middle of the photograph is the cooling pond.

Courtesy of Associated Press

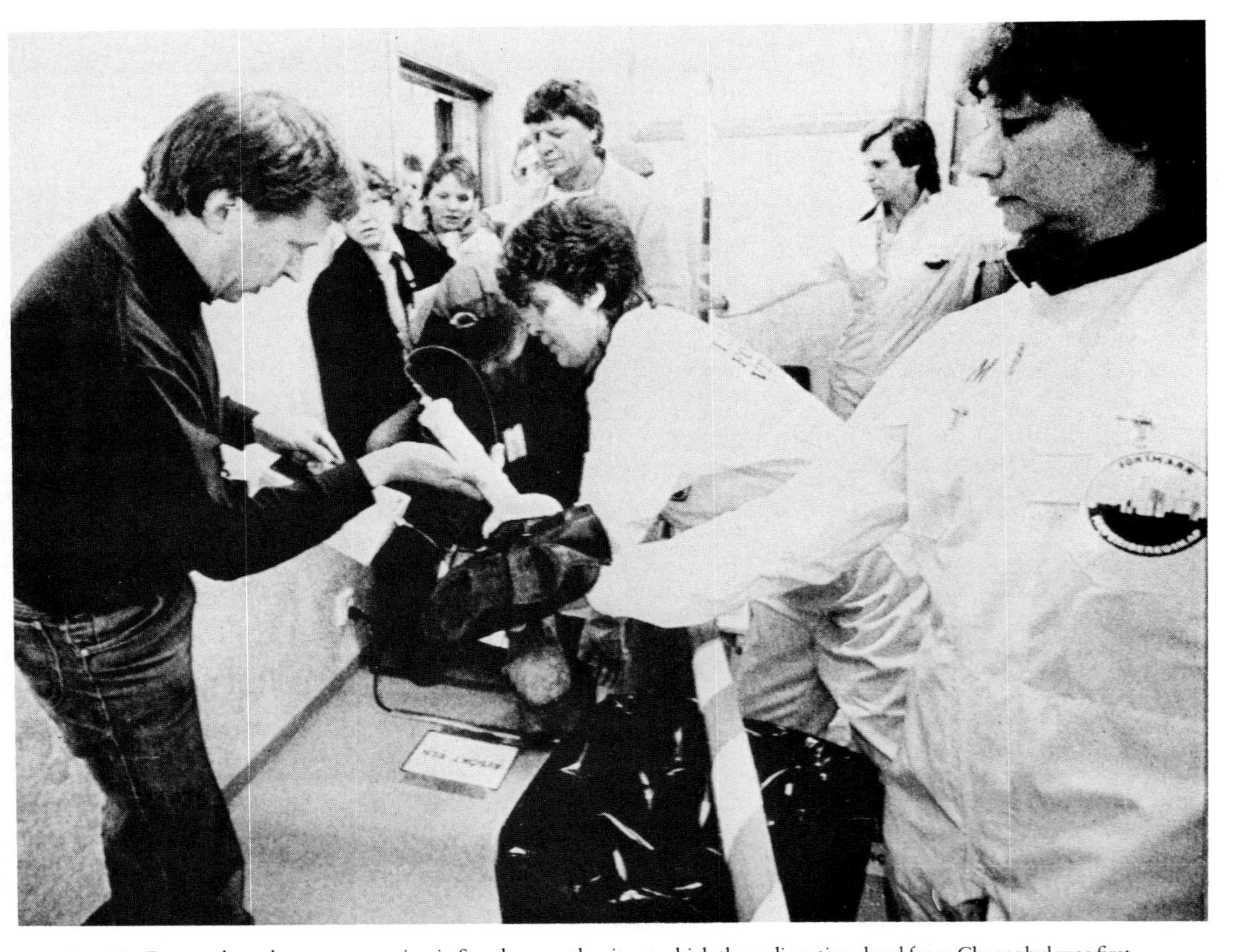

FIG. 33. Forsmark nuclear power station in Sweden was the site at which the radioactive cloud from Chernobyl was first detected outside the borders of the Soviet Union. This photograph was taken on 28 April 1986 and shows villagers from near Forsmark being monitored for contamination.

Courtesy of Popperphoto

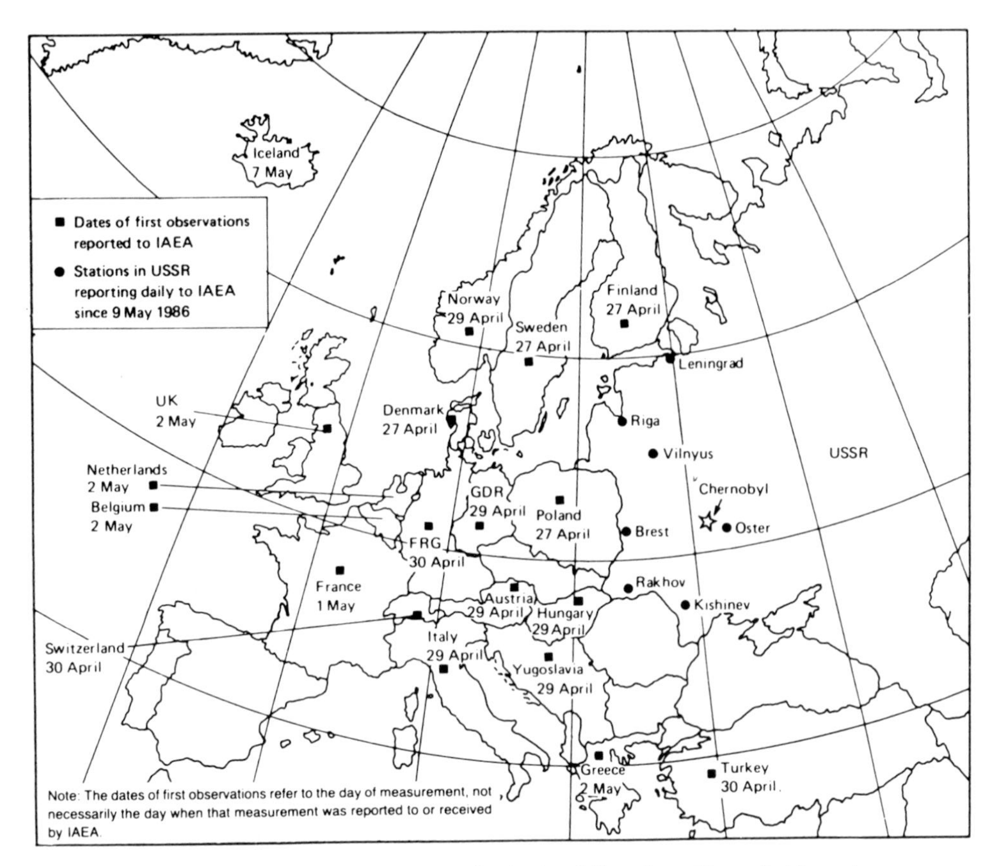

FIG. 34. Dates of first measurement of radiation fallout from Chernobyl in various countries, 27 April to 2 May 1986.

Courtesy of IAEA

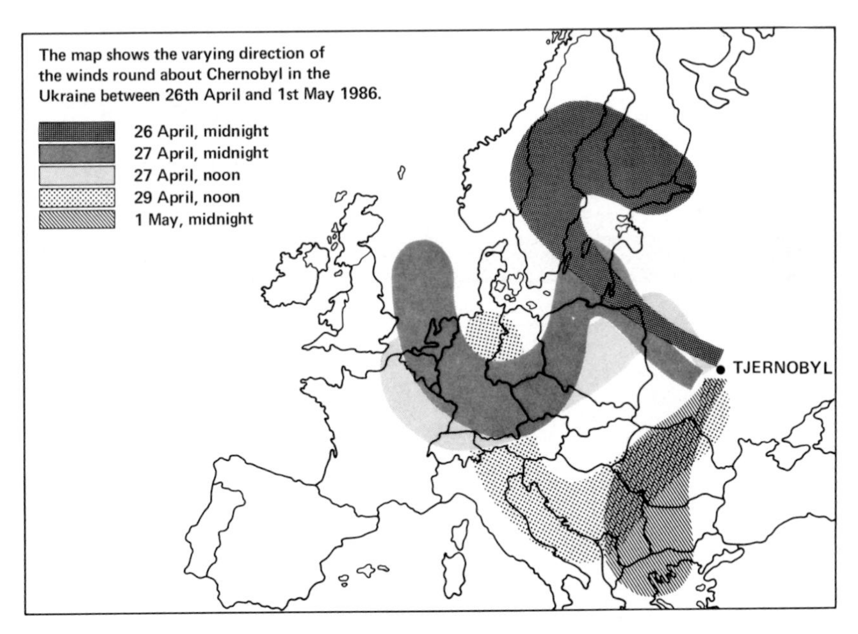

FIG. 35. The varying direction of the winds about Chernobyl between 26 April and 1 May 1986.

After map published by Swedish authorities in a special issue of News & Views: Information for immigrants

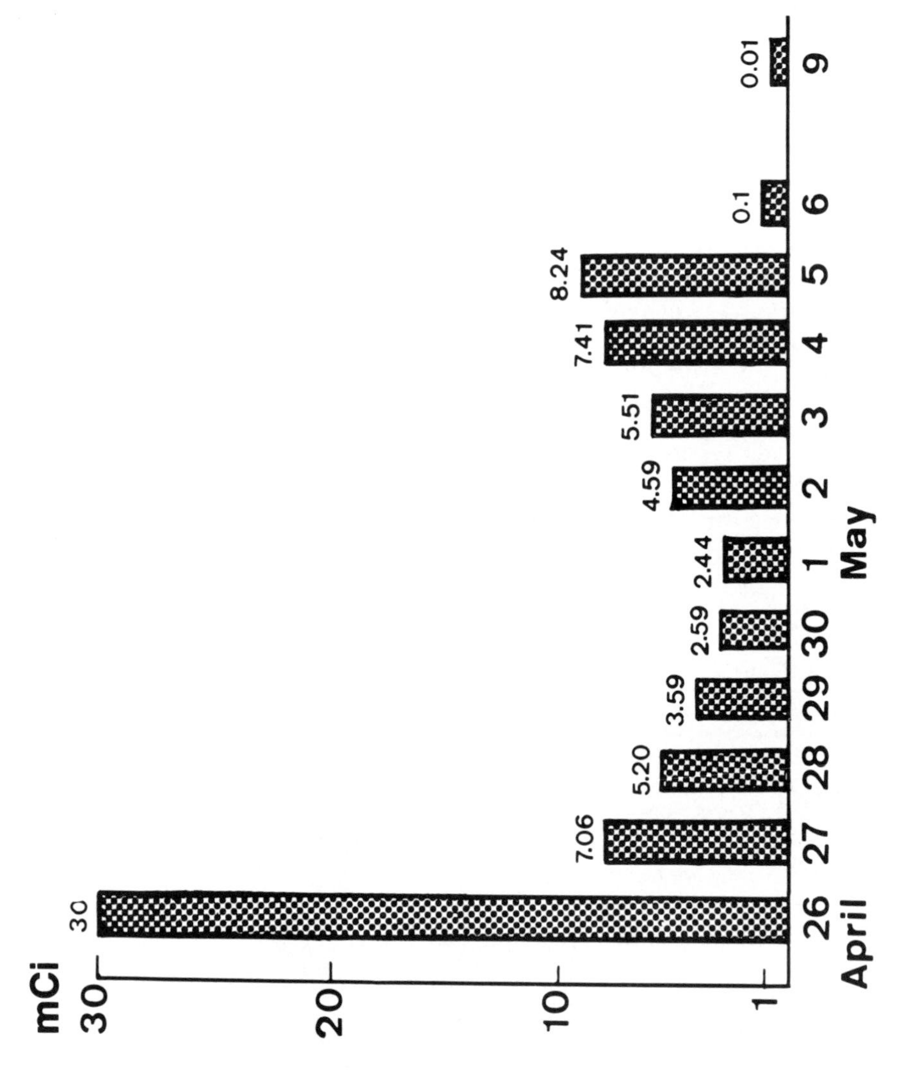

FIG. 36. Pattern of radioactive releases (in millicuries) from the Chernobyl reactor.

FIG. 37. "Is this a radioactive cloud over Kiev?" The photograph published by the *Sunday Times* and by *Time Magazine*.

Courtesy of The John Hillelson Agency Limited

FIG. 39. "Are these the fearful Kiev students?" The photograph was published by the *Sunday Times*, but the original Associated Press caption was "Kiev residents relax in the sun". This photograph was taken by Boris Yurchenko of AP during a government-escorted tour for representatives of Western news organisations on 8 May 1986.

FIG. 40. A young Kiev resident reads the labour union newspaper *Trud* on 8 May 1986.

Courtesy of Associated Press

FIG. 41. Kievskaya Gorilka.

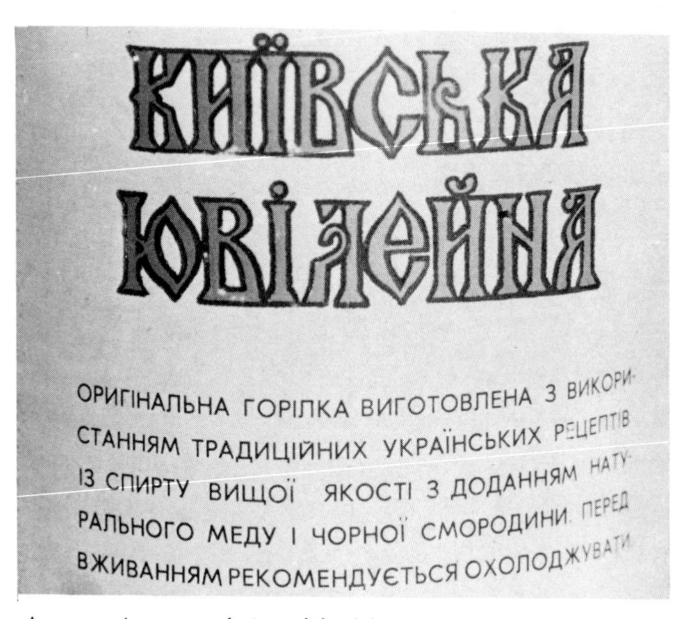

An approximate translation of this label is:

Original firewater prepared according to selected traditional Ukrainian recipes from spirit of the highest quality with addition of natural honey and blackcurrant. Refrigeration is recommended before use.

FIG. 42. Line drawing of photograph published in *Pravda* on 19 May 1986. It shows radiation scientists monitoring inside the nuclear power station, and clearly illustrates the protective clothing which includes gloves, overshoes, face mask and overalls.

FIG. 43. The corridor connecting unit No. 3 and the damaged unit No. 4 was a high radiation zone after the accident and workers were recommended to run through such areas. This photograph, taken May 1986, shows (*right in white overalls*) Aleksandr Yurchenko, head of a radiation monitoring squad, and (*left*) his assistant, Valeri Starodumov.

Courtesy of Novosti

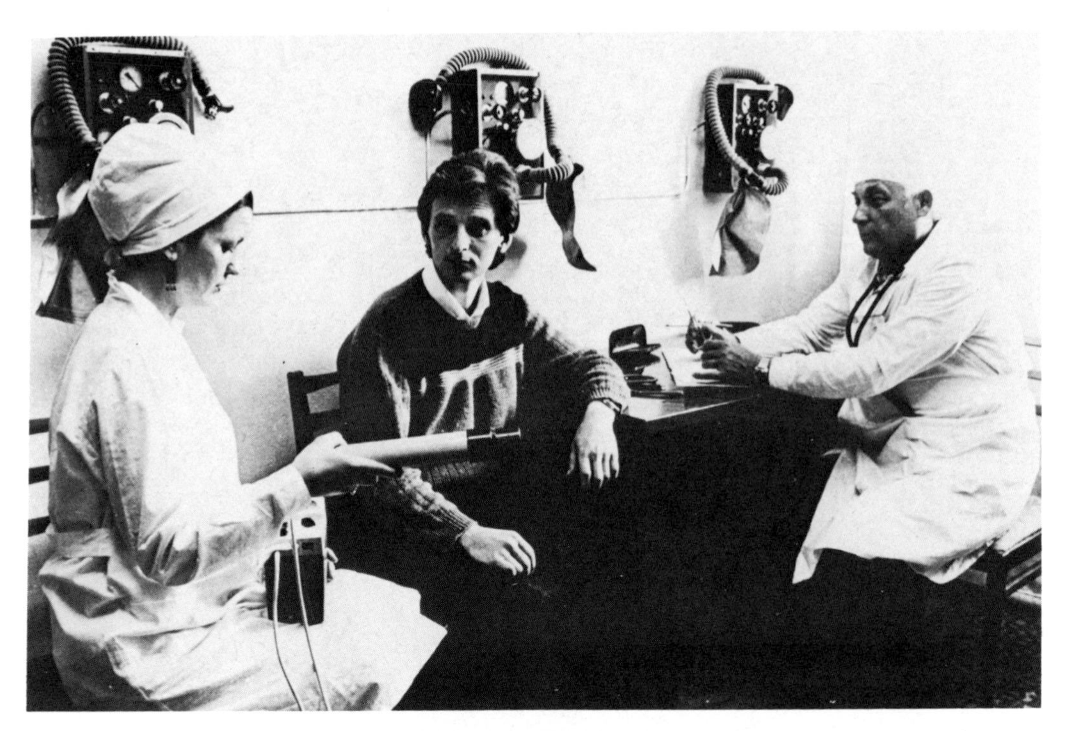

FIG. 44. An engineer from the Chernobyl power plant is checked for radioactive contamination by a health physicist (*left*) whilst a doctor looks on. This engineer, V. Taranov, was one of 150 duty personnel working on the three still intact units, May 1986. Most of the radiation victims on site at the time of the accident were young males, but at least two females were irradiated. One was a 58-year-old plant guard (see Case III in Chapter 4, page 70) who developed severe late skin lesions and died of a cerebral vascular accident. The occupation of the other female who was aged 63 is not known, but she received an estimated total body gamma radiation dose of 7–10 Gy was treated by foetal liver cell transplant and died 30 days post-accident.

Courtesy of TASS

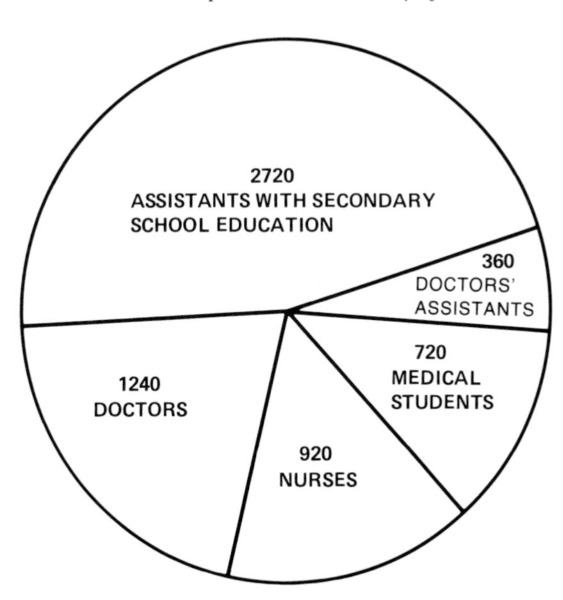

FIG. 45. Distribution of the total personnel of 5960 in the 450 medical brigades formed in the first few days after 26 April 1986. Almost half were termed "assistants with secondary school education".

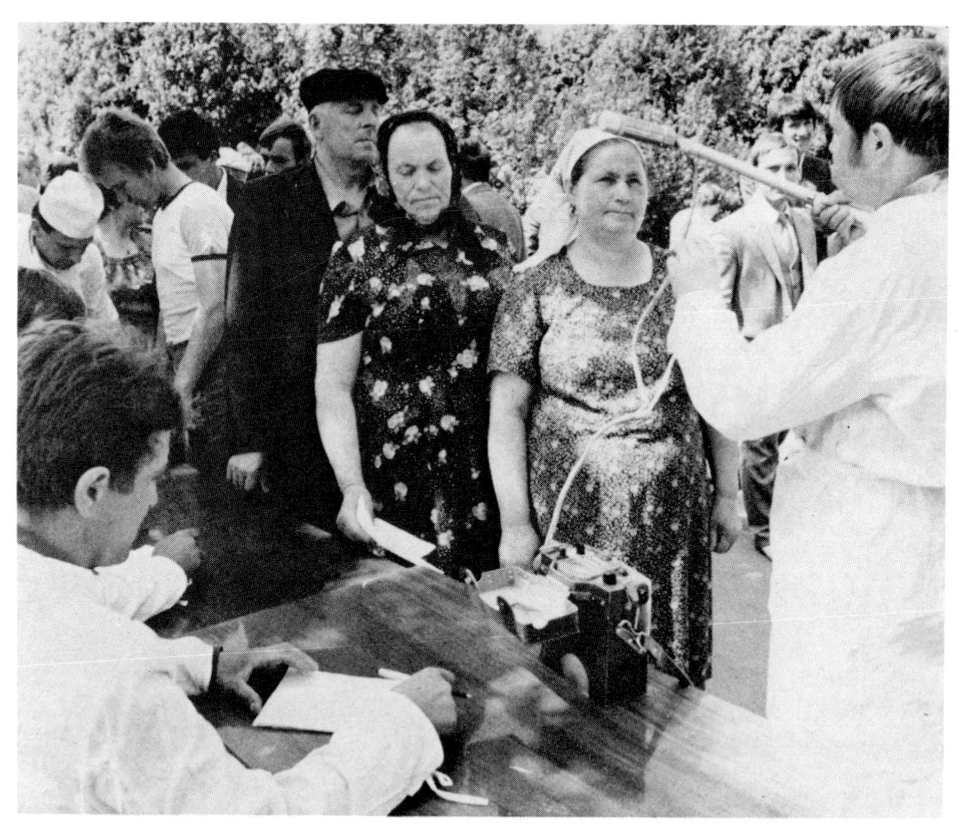

FIG. 46. No pictures from the Soviet Union relating to foodstuffs, including milk, or to stable iodine administration, were located at any photographic news agency. All that were available were photographs of monitoring procedures of the evacuees. This photograph shows a medical technician checking "body radiation levels" of Chernobyl evacuees moved to the Kopelova state farm near Kiev. It is not known how sensitive is the radiation monitor, but such a device would probably only be useful for checking external contamination on clothing, or relatively high iodine-131 uptakes in the thyroid. For a gamma ray spectrometry analysis of body intake, a special whole body counter is required, such as that used for some of the 203 cases diagnosed with a radiation syndrome, 11 May 1986.

Courtesy of Associated Press

FIG. 47. A traffic militia policeman at a vehicle check-point on the boundary of the restricted zone, 13 June 1986.

Courtesy of Popperphoto

Fig. 48. Radiation monitoring at Vienna Airport of some of the 64 passengers arriving on board a special Austrian Airlines flight from Minsk, 1 May 1986. Most of the passengers were relatives of Austrians working in the Soviet Union.

Courtesy of Associated Press

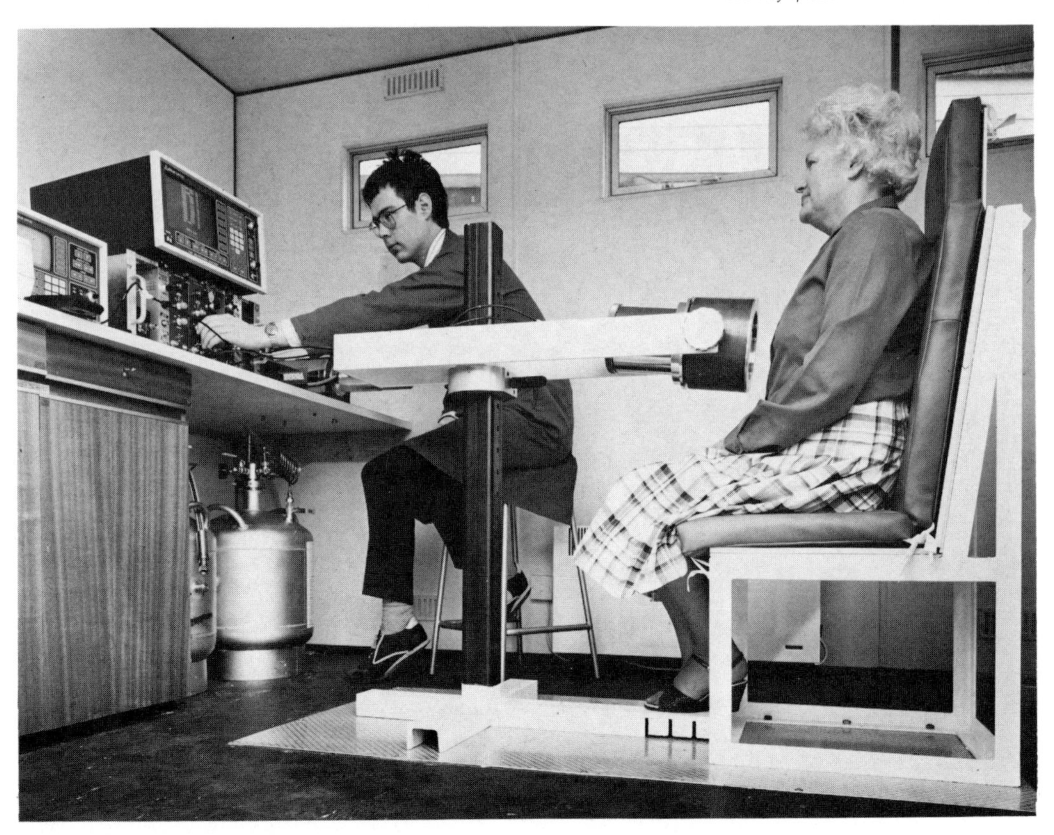

Fig. 49. Caravan containing whole body monitoring equipment used by the National Radiological Protection Board at London's Heathrow Airport.

Courtesy of NRPB

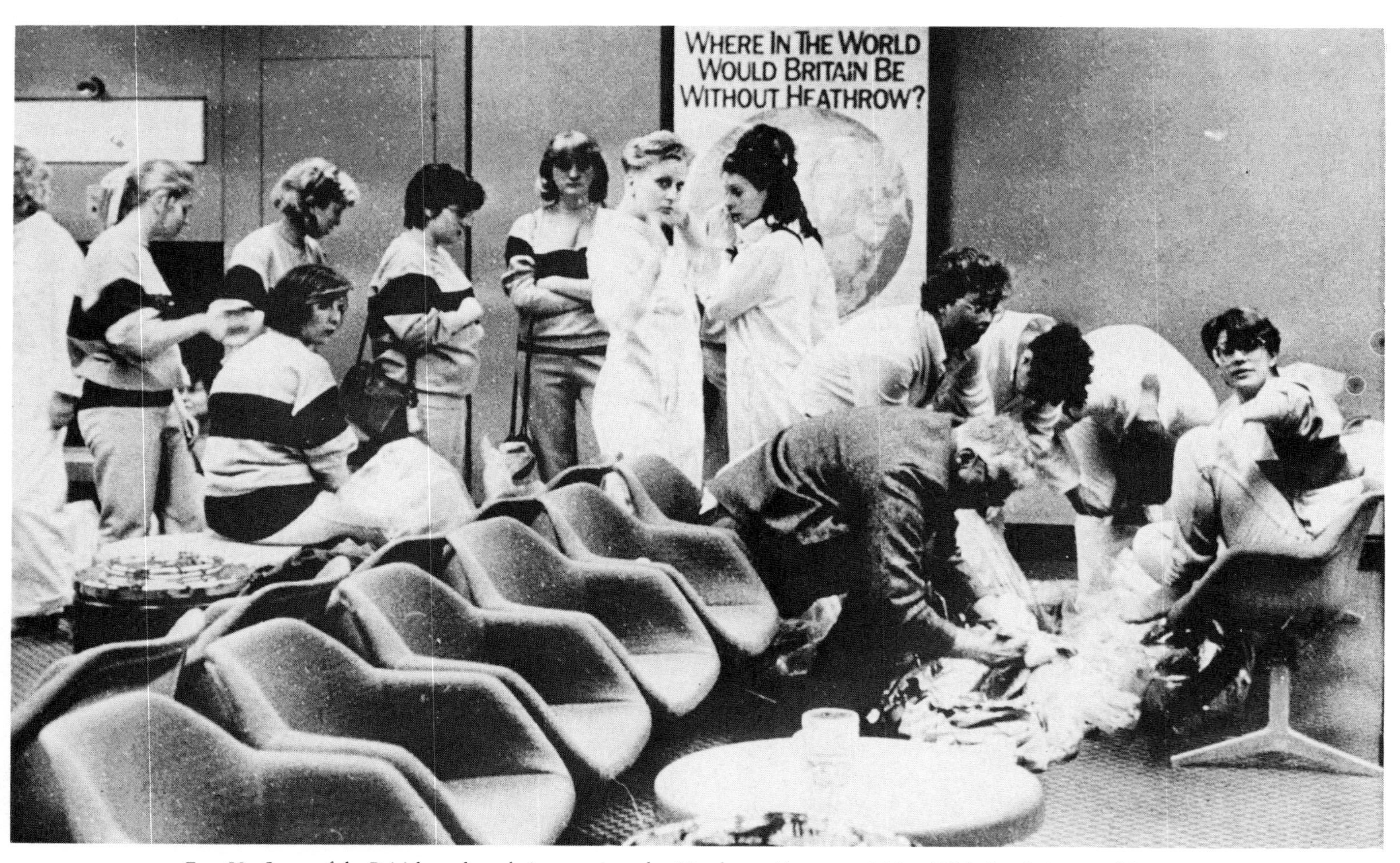

FIG. 50. Some of the British students being monitored at Heathrow Airport on 1 May 1986 after their arrival from Kiev and Minsk. Some of the students were given clothing in the Soviet Union because their own had been contaminated.

Courtesy of Popperphoto

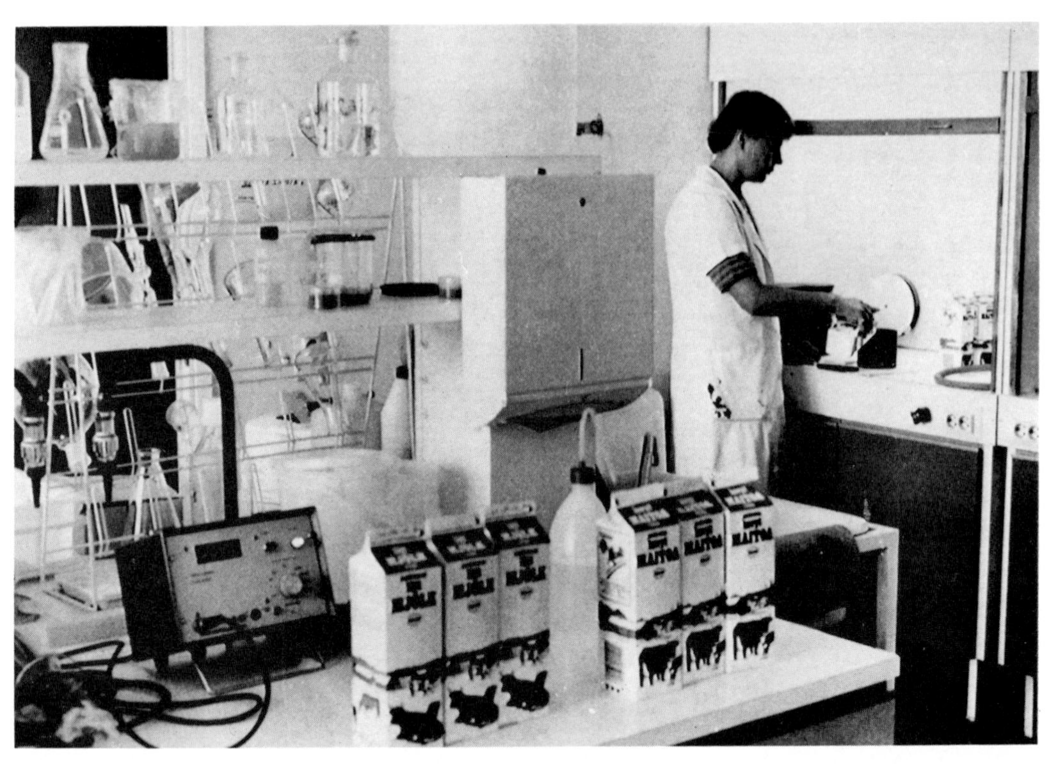

FIG. 51. Milk monitoring laboratory at the Finnish Office of Nuclear Radiation Safety, Helsinki, 30 April 1986.

Courtesy of Associated Press

FIG. 52. Blood tests on living reindeer to determine the contamination from caesium-137, August 1986.

Courtesy of Pressensbild

FIG. 53. Workers of Milan's fruit and vegetable market, one of the largest in Italy, dump all fresh vegetables, 9 May 1986.

Courtesy of Associated Press

FIG. 54. Drinking pasteurised milk in Zurich, Switzerland, 12 May 1986. Some 150 parents and their children had been demonstrating outside the cantonal parliament against the use of nuclear power.

Courtesy of Associated Press

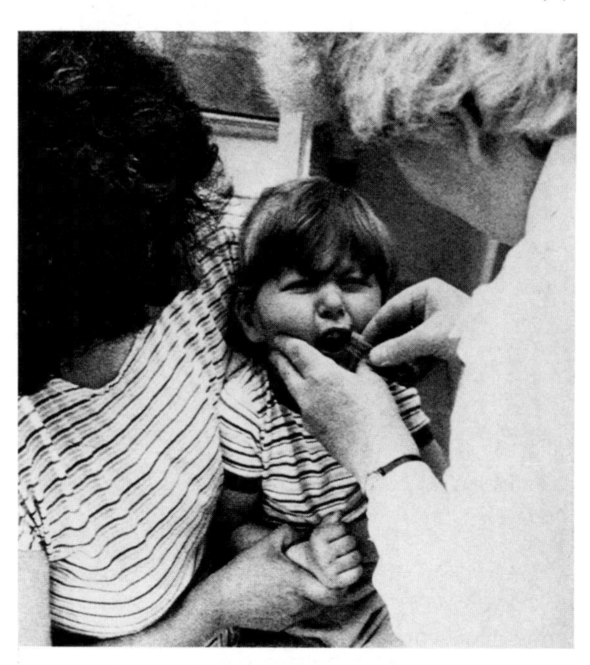

FIG. 55. Stable iodine preparations, tablets or solutions, are given to block the thyroid before any exposure to radioactive iodine-131. In this photograph a 3-year-old is held by her mother while the iodine is given, Warsaw, 30 April 1986.

Courtesy of Associated Press

Fig. 56. Radioactive warning sign in a pasture in Michelstadt, near Frankfurt, in the Federal Republic of Germany. Many farmers ignored the warnings that livestock should be kept in stables, 5 May 1986.

Courtesy of Associated Press

Fig. 57. A Greek anti-nuclear demonstration at the time of Chernobyl. Some 10,000 people took part in Athens, 13 May 1986, and those dressed as skeletons poured bottles of milk in the street.

Courtesy of Associated Press

ШЕРЕНГА НОМЕР ОДИН

Специальный корреспондент «Известий» Андрей ИЛЛЕШ передает из района Чернобыльской АЭС

FIG. 58. The 19 May 1986 edition of *Izvestia* contained photographs of the six firemen who died and eyewitness accounts from their colleagues. The firemen who died were (*from top left to bottom right*) Sergeant Nikolai Vasilievich Vashchuk, Senior Sergeant Vasilii Ivanovich Ignatenko, Lieutenant Victor Nikolaevich Kibenok, Lieutenant Vladimir Pavlovich Pravik, Senior Sergeant Nikolai Ivanovich Titenok and Sergeant Vladimir Ivanovich Tishchura. The heading of the article reads: RANK NUMBER ONE "*Izvestia*'s special correspondent, Andrei Illesh, reports from the scene of the Chernobyl Nuclear Power Station".

Ivan Mikhaelovich Shavrei

"... *A. Petrovskii and I went up onto the roof of the machine room; on the way we met the kids from the Specialised Military Fire Brigade No. 6; they were in a bad way. We helped them to the fire ladder, then made our way towards the centre of the fire where we were to the end, until we had extinguished the fire on the roof. After finishing the job we went back down, where the ambulance picked us up. We, too, were in a bad way* ..."

V.A. Prishchepa

"... *I climbed onto the roof of the machine room by the fire escape. When I climbed onto it, I saw that the roof's overhead covers had been destroyed. Some had fallen through, others had become loose.... In the morning I became ill. We washed and I went to the medical centre.... I don't have any more details.*"

Private Andrei Nikolaevich Polovinkin

"... *We arrived at the scene of the accident in 3–5 minutes. We started to turn the fire engine and to prepare for extinguishing.... I went onto the roof of the generator twice to pass on the brigade leader's order: how to deal with it. I personally would like to place on record a favourable mention of Lieutenant Pravik who knowing that he had received severe radiation burns still went and found out everything down to the smallest detail....*"

Sergeant Aleksandr Petrovskii

"... *Ivan Shavrei and I were ordered to go up the outside staircase to liquidate the fire on the roof. We were there for 15–20 minutes. We put out the fire. Then we went down: it was no longer possible to stay there. About 5–10 minutes later the ambulance collected us. That's all.*"

I.A. Butrimenko, Section Commander

"... *In such a situation no one could allow himself any weaknesses....*"

Courtesy of Izvestia

FIGS. 59, 60 & 61. Radiation injuries to firemen who fought the blaze at Chernobyl. These illustrations, or even similar ones, have not yet been released to either TASS or Novosti in the United Kingdom, have not yet been published in *Pravda* or *Izvestia*, nor, as far as I am aware, in any medical literature. Obvious signs of radiation induced skin injuries did not develop until 3 days after the accident, when a transient skin erythema developed which lasted no longer than 1 day. One group of patients then developed widespread erythema 5–10 days later with areas of skin breaking down, some of which required surgery. It is thought that this might have been caused by non-uniform exposure greater than 80 Gy to these areas of skin (some of which were exposed and some of which were covered by clothing). Another group of victims developed a moderate to severe reaction 21–24 days after the accident and these particular skin reactions were similar to those which can develop after radiotherapy and therefore suggest surface skin doses in the range 20–80 Gy. The most severely affected patients were those who had remained in the area of the accident for up to 5 hours and in these cases their clothing had become sufficiently contaminated to expose covered skins areas to a radiation reaction. In addition, some 28–30 patients also experienced a late wave of radiation erythema 2–4 months after the accident. This late stage occurred after the acute skin reactions had healed and recovery had occurred from bone marrow damage.

Displayed by the Soviet delegation as slides, IAEA, Vienna, August 1986

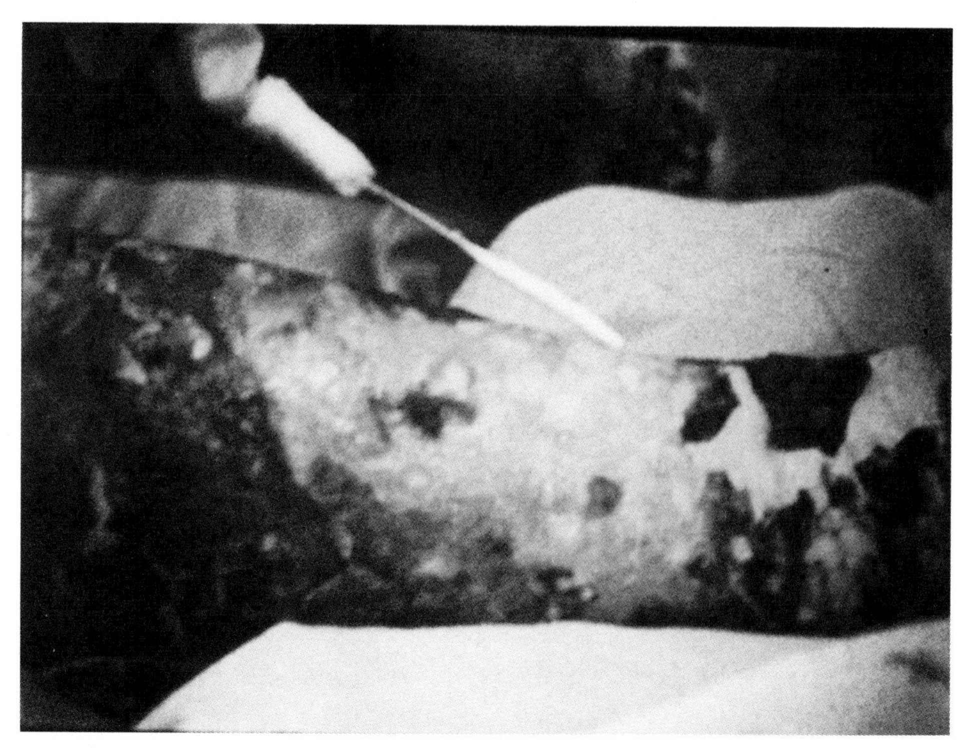

Fig. 59. Displayed by the Soviet Delegation as slides, IAEA, Vienna, August 1986.

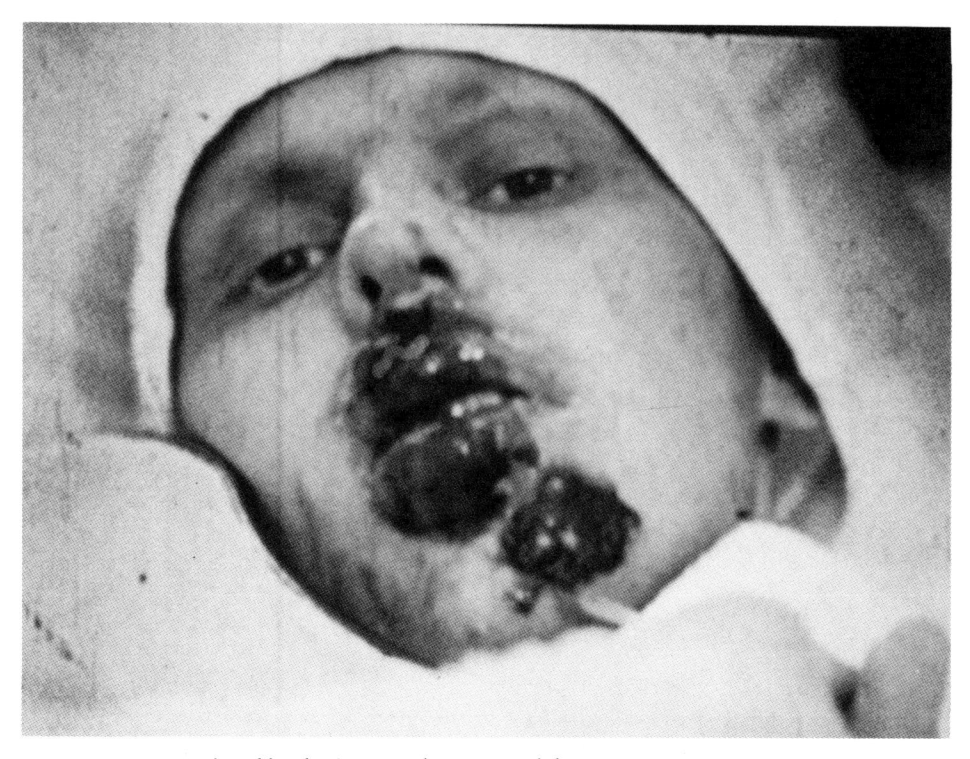

Fig. 60. Displayed by the Soviet Delegation as slides, IAEA, Vienna, August 1986. Dr Nadezhina later stated that the viral infections were of a herpes type (*Herpes simplex* of the facial skin, lips and oral mucosa) and were treated with Acylovir.

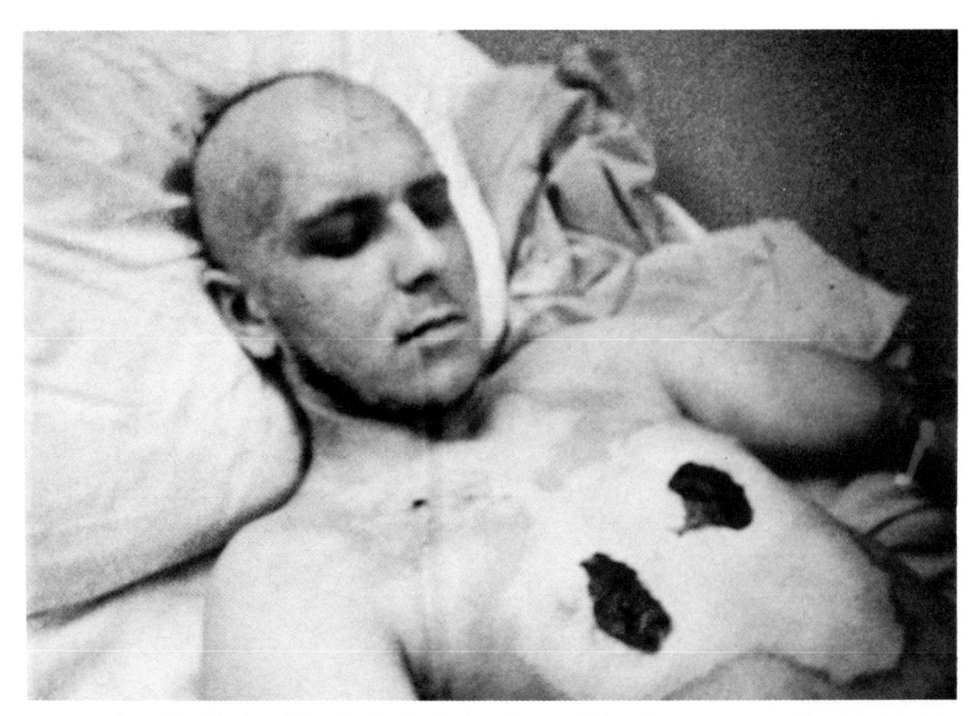

Fig. 61. Displayed by the Soviet Delegation as slides, IAEA, Vienna, August 1986.

Fig. 62. Two elderly people evacuated from the Kopelovo state farm, 50 kilometres west of Kiev. In all, about 1000 people were evacuated from the farm. The sign on the building in cyrillic letters reads "Evacuation Point". Prior to the evacuation a list of precautions were set up at the farm which included instructions regarding a limit to children's playtime outside in the open and a caution about dust on leaves, as well as asking that children be kept away from grass and trees.

Courtesy of Associated Press

4

The Victims

"Life, vain gift of chance, pray tell me—
Why have you been granted me?
Why, O fate, do you compel me
To endure such agony?"
(Alexander Pushkin, 1799–1837)
(On the occasion of the poet's 29th birthday, 26 May 1828)

THE THIRTY-ONE fatalities reported in the 15 November 1986 issue of *Pravda*, when the reactor was stated to have finally been safely buried in reinforced concrete using automatic machines, were not the only victims of the Chernobyl accident. The 135,000 evacuees from the 30-kilometre zone and the people who eventually will be included in the number of excess cancer deaths statistics are also victims of the disaster, as I suspect will be some national nuclear power plant programmes for the production of electricity. However, without doubt the major victims are those who have already died, most of them members of the firefighting brigades, and those who are still alive but who have experienced some degree of radiation syndrome.

Along with a number of the firefighters, some of the plant's emergency personnel received high radiation doses (above 100 rem) and suffered burns. Only five persons received significant thermal burns, but there were many more cases of beta-radiation burns. By 0600 on 26 April, one plant worker had died of severe burns and 108 people had been hospitalised, with a further twenty-four being later admitted to hospital. One of the plant workers was never found and remains buried in the entombed reactor. Up to 10 May, several hundred thousand people were medically examined, including blood tests, and 299* were diagnosed as having radiation syndrome, but these cases were confined to firemen and plant workers and there were none in the general population. This should be noted, because in the immediate days following world press coverage of events at Chernobyl there were recorded cases of people diagnosing themselves as having radiation sickness when all they had was a stomach upset. More unusual stories included the vegetarian who survived

* At the news conference on the effects of the accident reported by TASS and held in Moscow on 22 April 1987, Academician Leonid Ilyin revised this figure down to 237. He reported that of these, 209 were cured, 196 could return to work but still had to follow restrictions about contact with radioactivity, and thirteen had become invalids, with some of these having to undergo surgery. Approximately 100,000 people had undergone comprehensive medical check-ups.

3 days by eating only peanuts, the lady who wanted to know if she should dodge rainspots because of radioactive fallout, and the newly returned visitor from Eastern Europe who wanted to know if he was radioactive because an Eastern European lady had recently been breathing heavily over him!

These and similar stories are the fault not only of some scaremongering press coverage, but also of a lack of preparedness by governments for speedy dissemination of disaster information to the general public, and the lack of a means of educating this public in a simple manner as to what are radiation hazards. Most governments have emergency plans for dealing with localised high-level radiation accidents, but not for dealing with generalised alarm in the population about low-level radiation risks. One of the resolutions following the accident should be to educate the general public, as far as possible, on radiation risks and benefits (the benefits of radiation treatment and diagnosis in medicine should not be forgotten). However, correct and clear language must be used for this difficult task. This is illustrated by the first of only two "official" remarks from the Soviet Union during the 25–29 August 1986 meeting which raised a certain amount of laughter. The first referred to the planning for the visit to the Chernobyl site of Dr Hans Blix, Director General of the IAEA. It was stated that some Soviet officials were against the idea of this visit and gave as their reason that they were "worried that the radiation might harm Dr Blix's organism". The second time a peal of laughter was heard was at the end of one of the daily press conferences when the nuclear engineer Professor A. A. Abagyan was asked how many persons were on duty in the control room and the turbine hall at the time of the accident. On being given the answer, which totalled about seven, the following exchange took place:

QUESTIONER: "What happened to them?"

ABAGYAN: "You mean medically?"

QUESTIONER: "No."

(*Pause.*)

ABAGYAN: "They have been punished."

QUESTIONER: "How?"

(*Even longer pause.*)

ABAGYAN: "I am not an expert in that field."

(*Press conference brought to an abrupt close.*)

The immediate initial medical assistance was given by the Chernobyl regional hospitals and institutes which serve the power plant. Very early on, broad clinical criteria had to be established to decide between those who were very sick and needed immediate hospitalisation (group 1), those who were not very sick (group 2) and those who had no symptoms of radiation syndrome (group 3). This grouping took into account both the time and commencement of the syndrome and its severity and, although of necessity this was only a roughly defined grouping, it was nevertheless found to be important in practice. The early symptoms which placed people in the immediate hospitalisation group 1 included the onset of vomiting during the first half an hour or so of work, diarrhoea, temperature increase, emissions from mucous membranes and hyperaemia of the skin, particularly on open surfaces. One hundred and thirty persons were found to be in this group and all were hospitalised locally by the end of the first day, with 129 later being sent to specialist hospitals in Moscow and Kiev. The 129 represented the majority of people (ultimately 203) found with acute radiation syndrome of the second, third and fourth (the most severe) degrees. Only a small number of people, compared to the 129, were later diagnosed with light injuries, and these were found within 1 week following the accident.

The Soviet doctors were convinced that, for a reasonably accurate future prognosis to be made in the first hours, there were two important medical pieces of information it was necessary to find from the blood analysis studies which could be repeated two or three times in the first 24 or 36 hours. The first was the total number of leucocytes and the other was the formula for the differential counting of the leucocytes. This would make it possible to diagnose those with lymphopenia. Priority hospitalisation cases for Moscow and Kiev were then flown out in the first 36-hour period using three aeroplanes assigned for the purpose. An initial diagnosis of the degree of radiation syndrome was made within the first 3 days, and although a further review was made later on, only a few alterations to the initial classifications were made. Some people were reallocated from second to third or from third to fourth, but very rarely was it the other way round.

Degree of Radiation Syndrome

Degree	*Radiation dose received* (100 rad = 1 Gy)
First	less than 100 rad
Second	100– 400 rad
Third	400– 600 rad
Fourth	600–1600 rad

The characteristics of the fourth degree were described by Professor Guskova (from the Moscow hospital which treated the victims) as follows: "The fourth degree is the worst degree and the period here is short, 6–7 days. Primary reactions are early, in the first 15–30 minutes. The number of lymphocytes is less than 100 per microlitre. On the 7th to 9th days you have vomiting, damage to the digestive tract and the granulocytes are less than 500 per microlitre. Thrombocytes from the 8th to 9th days are less than 40,000. General intoxication is clearly shown and there is fever. In eighteen cases there were great beta-radiation burns on large areas of skin and in two patients there were also heavy thermal burns. The lethal outcome commenced from the 9th day, twenty-one patients with fourth degree radiation syndrome were all dead by the 28th day." Confirmation of the diagnosis of the less severe first degree of radiation syndrome required a much longer observation period for the patient, some 1 to 1½ months.

Once the acute radiation syndrome patients were in the specialised hospitals, repeated investigations of the bone marrow and peripheral blood were made, and radiation dose assessments were derived from chromosome aberration measurements. The following two tables are based on slides presented by Professor Guskova on 25–29 August 1986, which were not included within the Soviet working document issued to delegates.*

Grouping of 203 Patients by Severity of Radiation Syndrome

Degree of syndrome	Number of patients in Kiev and in Moscow		Deaths	Radiation dose (Gy)
Fourth	2	20	21	>6–16
Third	2	21	7	>4– 6
Second	10	43	1	>2– 4
First	74	31	0	>1– 2

Time of Deaths

The data presented at the August 1986 IAEA Post-Accident Review meeting and elsewhere was revised in the Guskova paper given at the 7–12 September 1987 ICRP meeting in Como, Italy.

Degree of syndrome	Time of deaths in days: August 1986 data	Time of deaths in days: September 1987 data
Fourth	9–28 (only a range was stated)	14, 14, 14, 15, 17, 17, 18, 18, 18, 20, 21, 23, 24, 24, 25, 30, 48, 86, 91 (for nineteen cases) +one case who died in Kiev on the 10th day from combined thermal and radiation injuries
Third	14–49	16, 18, 21, 23, 32, 34, 48 (for seven cases)
Second	Not stated	96 (for the only case)

Note 1. The total number of deaths (September 1987 data) in the above table is twenty-eight, but in addition to these, one other person died at the power plant (a reactor operator, Valery Khodemchuk, whose body was never recovered and is entombed with the radioactive debris) and one died during the first 12 hours after the accident (a power plant worker Vladimir Shashnok), from thermal burns, in the power plant's local Pripyat hospital where he was given first aid. This totals thirty victims. The total given in August 1986 was twenty-nine (see above). However, the figure most often now quoted in the press is thirty-one, the "extra one" presumably a fourth degree victim.
Note 2. Not all patients examined experienced radiation sickness, and in the group that did not, only one-half showed signs of chromosome aberrations. The dose maximum for these patients was estimated to be in the range 0.2–0.8 Gy.
Note 3. The indication used for transplantation of allogenic bone marrow (TABM) and transplantation of human embryonic liver cells (THELC) was a whole body gamma-radiation dose estimated to be of the order of 6.0 Gy or higher. TABM was undertaken thirty times and THELC was carried out six times. After THELC, all patients died in the early stages (14–18 days post-irradiation) from lesions of the skin and intestine. The only exception to this was a woman aged 63 years who survived 30 days post-irradiation after receiving a dose of 7–10 Gy. Her death occurred 17 days post-THELC. Seven patients died at 2–19 days post-TABM (15–25 days post-irradiation) from acute radiation lesions of the skin, intestine and lungs. Of six patients who had no lesions of the skin and intestine, which were considered to be incompatible with life, and whose doses were estimated to be 4.3–10.7 Gy, two survived after TABM (their doses were 5.8 and 10.7 Gy). Both their donors were the patient's sisters. Four patients died at 27–79 days post-TABM from mixed viral–bacterial infections (two [5.0–7.9 Gy and 5.8–6.0 Gy] against a background of a well-functioning graft and two [4.3 and 10.7 Gy] after early rejection).

* On 11 April 1987 the Institute of Biology held a seminar entitled "The Lessons of Chernobyl". Professor Guskova gave a presentation which I attended in which she reported that no further deaths had occurred and that changes to the eyes of the most heavily irradiated persons had been studied with a 1-year follow-up, confirming that no cataracts had so far been detected. I was also able to find out from her the occupations of those who had died. Only six of the fatalities were firemen, the rest being engineers, technicians, operators and other power plant personnel. None of the dead were medical staff or members of the general public. At the same meeting, Dr Robert Gale told of one survivor who was a cook at the power plant. A resident of Pripyat, she had received radiation skin burns to ler legs below the knees after walking to work through a field covered in high grass on 26 April 1986.

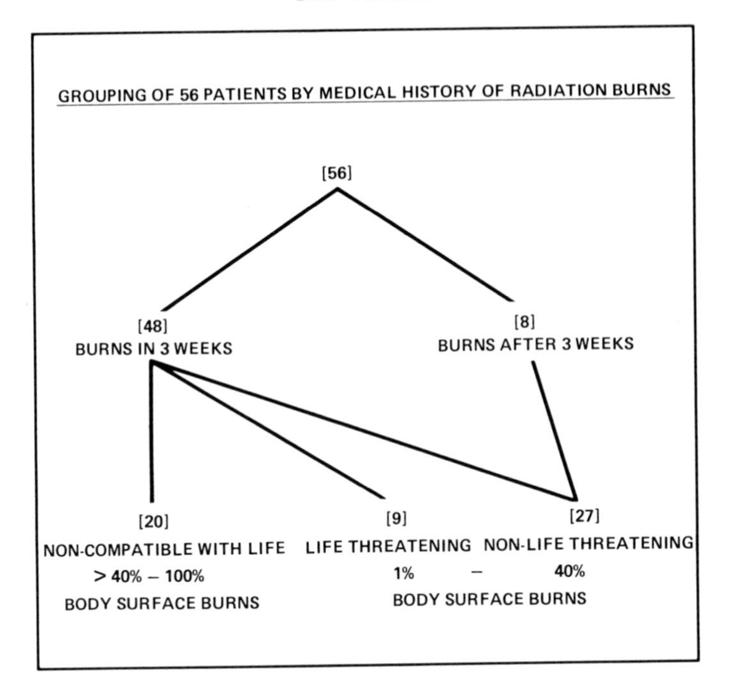

Other measurements were also made, both externally and internally, before and after decontaminating the patients as much as possible. These included the use of thyroid counting for radioactive iodine, gamma-ray spectrometry using whole body counters, urine analyses for radioactive content, biochemical and metabolic studies. The main gamma-radiation emitting radioactive isotopes found using the whole body counter were reported (USSR Delegation, 25–29 August 1986) to be iodine-131, iodine-132, caesium-134, caesium-137, niobium-95, cerium-144, ruthenium-103 and ruthenium-106. The total alpha radioactivity of plutonium radionuclides was estimated from the urine measurements and was found to be very low. The Soviet doctors considered that this was not due to the Chernobyl accident but to previous working environments. It was also found that no patients had received neutron irradiation.

Most of the terrible burns received were beta-radiation burns rather than thermal burns, although two patients in particular did have heavy thermal burns to the skin. This led to the entry both of dirt and of radioactive contamination through the open wounds and made any total decontamination impossible. The chart for the group of fifty-six patients represents all those with greater than 1% body burns, and they have been divided into two main groups according to when the burns appeared. In the larger group of forty-eight, burns were from several hours to 3 weeks, but in the smaller group of eight, the burns first appeared in the 4th, 5th, 6th or even 7th week after the accident. However, they were all what Professor Guskova termed "non-life-threatening burns".

In the group of forty-eight there were some cases which tragically experienced a second wave of radiation burns which covered new portions of their skin and extensions to earlier damaged skin. This was presumably because not all the radioactive particles could be decontaminated. They worked their way into the tissues during the decontamination procedure, and then produced the second wave of burns. If the burns covered more than 40% of the body surface, then the patient would inevitably die, whereas some patients even with less than 40% body surface burns were classified as being in a life-threatening state. The twenty-nine deaths in the chart correspond to the twenty plus nine more of the group of forty-eight patients.*

If the colour videotape showing the reactor's burning graphite moderator was dramatic, then so also were a series of some six or seven slides, shown by Professor Guskova, of the firemen victims at various stages in their medical history. Not all the pictures are reproduced, since the original slides were returned directly to Moscow on 29 August and those in this book (Figs. 59, 60, 61) are only photographs of some of the projected Guskova slides which were fortunately obtained by a colleague. But Professor Guskova's commentary makes it very clear what suffering must have been experienced by the firemen and the power plant staff who heroically remained to complete damage control to the best of their ability while knowing that they must be irradiated to levels far above those considered safe. The widespread extent of the beta-radiation burns, however, could not have been foreseen, since this was a totally unexpected feature of this accident. One practical point which arose from the slide presentation was that at least some of the radiation burns received by the unfortunate firemen might have been avoided. Radioactive particles were able to get down their necks in the gaps between their skin/shirt (the latter would have provided no protection) and the collar of their coat/uniform jacket. This could have been prevented if there had been a better design of fireman's helmet with some form of protection at the back, rather like the *kepi* of the French Foreign Legion.

The Guskova slide captions, as near as my tape recordings show, were as follows.

"Epilation of the *scalp*, and blue skin (Fig. 61), where there was total ulceration."

"This is a heavy burn. It is the *thigh* (Fig. 59), which has very deep damage. There are very painful sections. You have the blue pigmented parts and scabs are forming."

"This is a less serious change; it is the *back of the neck*. There are tendencies for the skin to open on the not very well-protected parts, or those parts where the cloth of the clothing is pushed in. This gives a sort of applied radiation burn effect."

"Patients have very many different kinds of burn. This one is another characteristic dark pigmentation. You have these waves which are moving into the *skull* of this patient, into the *ear*, in the *edges of the eye*. . . ."

---

* As part of the follow-up to the accident, the IAEA held an Advisory Group meeting in Paris on 28 September to 2 October 1987, entitled "Medical Handling of Skin Lesions following High Level Accidental Irradiation". The aim of the meeting was described by the IAEA in the following terms: "The accident at Chernobyl caused an unexpectedly high frequency of skin lesions in various combinations, i.e. thermal and chemical burns with contamination, or beta burns with thermal and chemical lesions, skin burns with various degrees of external irradiation, which is of great interest to radiopathologists, skin pathologists, oncologists and radiation protection specialists. New methods have been applied recently in the treatment of skin lesions, for example the use of artificially produced human skin which has enabled brilliant successes even in cases in which the area involved in the burns is 90% of the body surface (a clinical condition traditionally associated, until very recently, with unfavourable prognosis). The main goal of the Advisory Group was to elaborate therapeutic schemes for the various clinical conditions of skin lesions deriving from high-level accidental irradiation."

"A picture of this patient at a later stage. This is characteristic of the later stage period where you have extensive portions with pigmentation covering more or less the *whole of the torso*. Then there is a new wave, a new explosion in these small portions, the surface of ulcers where haemorrhages have occurred."

"This is characteristic of *skin* where you have *infection* as well (Fig. 60). This has spread to the membrane of the nose and it causes the patient the most terrible pain. Explosions of this sickness from viruses is something we had in about twenty patients."

An additional four case histories were described by Soviet doctors at the 28 September–2 October 1987 meeting in Paris on "Medical Handling of Skin Lesions following High Level Accidental Irradiation". They are given below* (Cases I–IV) and were selected to illustrate the range of problems that arose, possibly a reflection of the local radiation dose and energy of the beta-emissions.

Case I

Male plant worker who received an estimated average total body dose of 9 Gy. This patient received a bone marrow transplant from a female donor. This was rejected, but his haematological status improved due to recovery of his own bone marrow. This indicated a highly non-uniform distribution of the dose. He developed skin lesions from 5 days after irradiation, eventually involving 40% of the body surface area. He showed epilation of the scalp and eyelashes, but the eyebrows were not affected. Lesions were severe over both buttocks as a result of his sitting on a contaminated surface. These areas of skin developed blisters and foci of ulceration, which required covering with free skin grafts taken from the patient's flank 2 months after the accident. These 0.8 mm thick free skin grafts took fairly satisfactorily and the patient was released from hospital 5 months after the accident. Small areas of necrosis developed in the grafted areas 7 months after exposure, but these had healed by 12 months when he was otherwise well, although aspermia persisted.

Case II

This male turbine-operator received an estimated whole body gamma dose of 2.0–2.5 Gy and thus the bone-marrow-related syndromes were only slight to moderate. He developed severe erythema and oedema of the skin in the second week after the accident and by the end of the third week had developed widespread erosion of the skin over the wrists and also of the trunk and thighs. He experienced a protracted and severe fever that appeared to be associated with the severity of the skin lesions. Topical therapies were tried, but he required surgical intervention on the 50th day after exposure. This involved removal of dead tissue from the ulcerated areas of skin on both wrists and grafting with skin taken from the patient's flank. The graft on the right had proved unsuccessful, either because of the severity of the injury to the deeper tissues (most likely cause) or because of radiation injury to the grafted skin. A pedicle flap taken from the anterior abdominal wall was successful in covering the ulcerated area, becoming established within 3 weeks. Although there was no evidence that the tendons were damaged, the patient now had only limited movement of the wrists and is unable to use his hands. He unexpectedly developed an

* I am indebted to Dr John Hopewell for providing me with these four case histories.

annular ulcer at the base of the fifth finger on the left hand in April 1987. The patient requested amputation of this digit because of the associated severe pain.

Case III

This 58-year-old female plant guard was on duty in a booth about 300–500 metres from the reactor site at the time of the accident. She ran several kilometres from the site and as a result had dry radioactive contamination (contaminated soil a dust) of her legs and shoes. She received an estimated bone marrow dose of 3 Gy from which she recovered. She experienced three waves of erythema over her thighs and lower legs; the third wave of erythema developed almost 3 months after the accident and was accompanied by oedema and severe pain. Severe lesions also developed later in the feet. The severity of these late developing skin lesions was accompanied by a deterioration in her general physical condition. She subsequently suffered a fatal cerebrovascular accident which was probably superimposed on a condition of a generalised radiation induced vascular damage.

Case IV

A patient showed evidence of a moist skin reaction over his chest wall and of both hands 18 days after the accident. The nature of his exposure could not be determined. Dose was estimated to be 80 Gy of soft beta-irradiation. This reaction resolved by 8 weeks with minimal medication, which included the application of antibiotic cream. There was no tissue scarring, although the area was depigmented. Late evidence of telangiectasis and dermal atrophy occurred 6 months after exposure. There is no evidence of the progression of these late lesions.

HIROSHIMA CASE HISTORIES

Not many eyewitness accounts have been published of victims of atomic bomb burns, and therefore for comparison with the Chernobyl experience described by Professor Guskova, the following three accounts from Hiroshima are included. The first two were recorded by John Hersey (1946), a journalist sent by *The New Yorker* in May 1946 (the bomb fell on Hiroshima on 6 August 1945) and the third was published by Anne Chisholm (1985) in her follow-up of young girls who survived the explosion.

1. A group of victims seen by the Rev. Mr Kiyoshi Tanimoto of the Hiroshima Methodist church:

   "... huge raw flesh burns ... skin slipped off in huge glove-like pieces ... burns were yellow at first, then red and swollen, with skin sloughed off, and finally, suppurated and smelly ...".

These victims did not survive.

2. A group of victims seen by Father Wilhelm Kleinsorge, a German Jesuit priest:

   "... faces wholly burned, their eye sockets were hollow, the fluid from their melted eyes had run down their cheeks (presumably from having faces upturned when the bomb exploded) ... mouths mere swollen, pus-covered wounds which they could not bear to stretch enough to admit the spout of a teapot for a drink of water (a large piece of grass was used as a straw) ...".

These victims did not survive.

3. Personal memory of a 13-year-old girl, Shigeko, who was in a street 1.6 kilometres from the hypocentre:

> ". . . whole face burned, . . . no eyebrows . . . mother had to pull my eyes open . . . my skin came off when they tried to remove my burnt clothes . . . 4 days later the burned skin was peeled off my face, it was all black, underneath was full of pus . . .".

Shigeko survived, scarred for life.

In terms of injuries, the experience of the Chernobyl victims has, in the media, sometimes been likened to the experience of victims of Hiroshima. This is obviously false in terms of the number of casualties, but also these few accounts show that however bad Chernobyl was in terms of the cost to individuals, the experience of many of the Hiroshima victims was much worse. Additionally, it should be noted that the large number of Hiroshima burns were of thermal and not beta-radiation fallout origin. They were of two types, flash burns from the direct heat of the bomb fireball and flame burns from burning clothes and ignited buildings. There are many differences between casualties from a civil nuclear power plant accident and from nuclear warfare.

Much has been written on the bone marrow transplants undertaken by Dr Robert Gale of the University of California, Los Angeles, USA, with Soviet doctors and with Dr Yair Reisner, an Israeli biophysicist from the Weizmann Institute of Science at Rehovat, near Tel Aviv, Israel. It would seem from the large amount of coverage in the media reports that these bone marrow transplants were a major factor in the treatment of the victims. This is not true, and, as one can see from an earlier chart, the radiation burn factor of greater than 40% of body surface would have in itself prevented such transplants being successful in these cases. None of the deaths were due exclusively to bone marrow failure. For example, Professor Guskova concluded that "for radiation accidents the proportion of patients in whom transplantation of allogenic bone marrow is absolutely indicated for beneficial treatment is very small"; that "with reversible breakdown of myelopoiesis caused by overall gamma-radiation doses of the order of 6 to 8 Gy, the transplant may take, but this taking will always have a negative effect in therapeutic terms and even endanger life as a result of the higher risk of secondary disease developing"; and that such "transplants had a very, very minor effect".

In practice, six bone marrow and three foetal liver transplants were carried out before Dr Gale's team arrived in Moscow and seven bone marrow and three foetal liver transplants took place with Dr Gale's assistance (*The Lancet*, 26 July 1986, page 212). The human embryo cells (foetal liver) were used for those patients where the radiation dose was particularly large. The results were not successful. The other transplants were performed using only relatives as donors and in a search for suitable donors for thirteen patients, 113 people were tested with complete matching being found for nine patients. For the remaining four, additional special transplant manipulations were necessary.

The most recent data on the survival of the victims and on the results of bone marrow transplantation was presented by Professor Guskova on 7–12 September 1987 at the Como meeting of an ICRP Committee.

The presentation by Professor Guskova included the following two points.

1. "The change in the total number of victims from 203 to 237, which was declared in November 1986, was due entirely to patients with acute radiation syndrome of the first degree of severity. There were thirty-one patients of this group

in the specialised station hospital (see table earlier in this chapter for patients in Moscow) and 109 in Kiev (previously given as seventy-four)."

2. In an emergency situation such as after Chernobyl, "the group of persons for whom bone marrow transplantation is indicated and may succeed is very limited".

Finally, to end this chapter on a less depressing note, not all of the irradiated firemen victims died, and in particular Major Leonid Telyatnikov, (see Fig. 134) the Chernobyl fire chief whose team was the first to tackle the blaze, is still alive, after treatment in Moscow at the specialist hospital No. 6, and after recuperating at a rehabilitation centre outside Moscow is still alive. One of the illustrations (Fig. 105) in this book shows him with some of his more fortunate colleagues at the sanatorium.

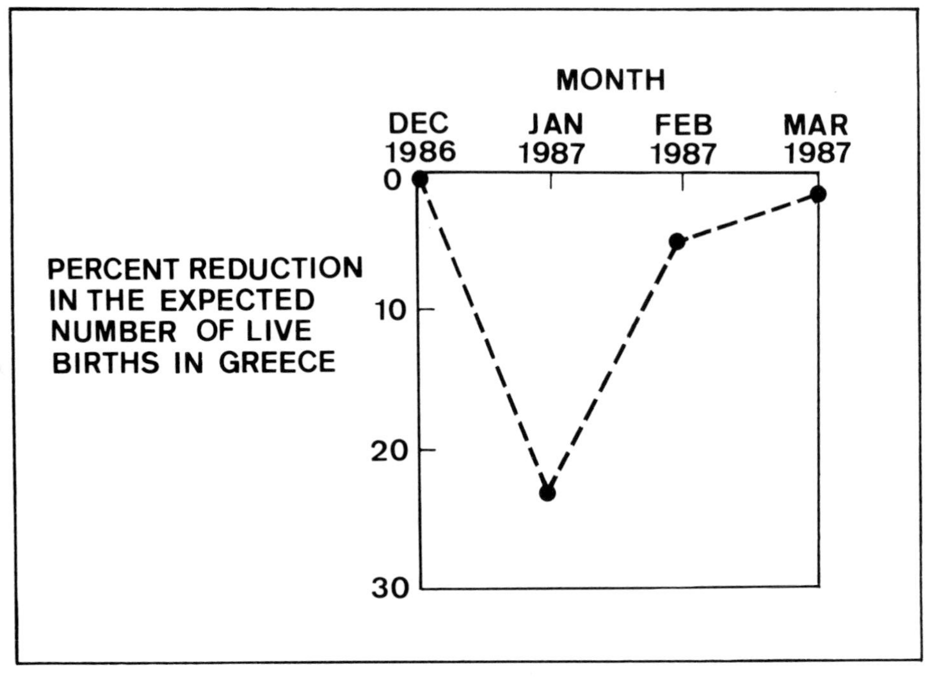

	JAN 1987	*FEB 1987*	*MAR 1987*
Observed no. of live births	*7032*	*7255*	*8350*
Expected no. of live births	*9103*	*7645*	*8453*

On 31 October 1987 a paper by Trichopoulos et al. was published on "The victims of Chernobyl in Greece: induced abortions after the accident". Women who had the first day of their last menstrual period during the month before the Chernobyl accident would be expected to give birth during January 1987. During most of May 1986 there was panic in Greece because of conflicting data and false rumours and many women thought they had a high risk of giving birth to an abnormal child. From the published data by Trichopoulos it is seen that 23% of early pregnancies at perceived risk were artificially terminated. By June 1986 the panic had subsided and the true radiation risks were better understood.

5

Evacuation

Total	135,000
Pripyat	49,000
Chernobyl	12,000

(USSR Delegation, 25–29 August 1986, Vienna)

THE SOUND track of the videotape from Soviet television which was shown in the exhibition area outside the IAEA conference halls, 25–29 August 1986, summarised the bare facts of the evacuation of Pripyat:*

"The evacuation of the town of Pripyat was announced at 1400 hours on 27 April. Although little time was allowed for packing up, the evacuation took place in an orderly fashion. People took only the bare essentials with them. 40,000 people left the town in 2 hours 45 minutes. Numerous tests of the air, water and the soil showed that the disabled reactor caused radioactive contamination of an area of about 1000 square kilometres around the plant. Some 116,000 people, the entire population, were evacuated from the affected area. Needless to say, children were the object of special concern. All the children were sent down to the southern Ukraine, the Crimea and the Black Sea coast. Within the shortest possible time, every condition was created for them to have a good rest. All the evacuees were given material assistance, money, clothes and houses. The misfortune brought people closer together. It demonstrated that we have a common cause, land and home. Old and young alike identified themselves with it. People offered their homes, clothes, money and blood to the victims. New villages and settlements were being built for countrified folk. A relief fund for the victims of the Chernobyl accident has been set up on grass roots initiative. Soviet people have donated more than half a billion roubles [1 rouble is approximately £1 Sterling] to bank account 904."

Prior to evacuation, which had been ordered when the radiation exposure rate in Pripyat reached 1 rontgen per hour, stable iodine had been distributed as a priority to all children's establishments in Pripyat, since they could be expected to drink more milk than adults, and since they also had smaller thyroids than adults (dose is absorbed energy per unit mass) they were more at risk from radioactive iodine-131.

---

* Chernobyl was evacuated 8–9 days after the accident. No reason was given on 25–29 August by the Soviet delegation, but it could have been logistical. The evacuation zone was originally 15 kilometres from the power plant, but was later extended to 30 kilometres, as reported in *Izvestia* on 7 May.

However, following monitoring measurements by Soviet experts, it was later estimated that 97% of the Pripyat evacuees receive a dose not greater than 30 rem, and that mortality from thyroid cancer might be increased by 1%. For the entire 30-kilometre evacuation zone, the Soviet estimates from measurements gave a maximum dose of 25 rem, with the proviso that a few people may have received 30–40 rem.

Panic by the inhabitants of Pripyat at the time of the evacuation was not mentioned in the Soviet TV soundtrack, but the authorities obviously had the possibility in mind when they planned the logistics for the transfer of some 40,000 persons at very short notice. At 0215 hours on 26 April there was a meeting of the Pripyat Internal Affairs Department, and their first decision was to bar all unnecessary traffic from the town. An *Izvestia* special correspondent reported on 7 May that their next priority was "to maintain order", and also that "no assembly points were established, so that commotion and panic could be prevented". Militia forces organised checkpoints, roadblocks, cordons and control point services. The most hazardous site was the checkpoint "near the enflamed building". The entire town of Pripyat was divided into five main sectors, each covering one housing estate, and five evacuation groups were manned accordingly. It took the night of 26 April and half of the next day to prepare lists of evacuees, and evacuation workers were distributed in accordance with the number of buildings and entrance doors. Eleven hundred buses were used, and these were distributed according to the five main sectors and given defined evacuation routes. For some reason (perhaps high levels of radioactive contamination?) the railway line system running through Pripyat was not used. Cattle were evacuated on 26 April, when some tens of thousands were loaded on to hundreds of open trucks. Military and civil authorities were involved in this operation, together with volunteers.

A *Pravda* report of 6 May describes the deserted town: "Only a specialised radiation monitoring motor vehicle appears from time to time. Also, periodically, the sound of engines disturb the silence of the settlement on the bank of the Pripyat river, a regular shift comes to the station, three units of the nuclear power station need supervision, and specialists are controlling the station's reactors that have been shut down", and "the Pripyat looks strange and unusual from a helicopter. Snow-white many-storied houses, wide avenues, parks and stadiums, playgrounds near preschool establishments, and shops. . . . A few days ago 45,000 people lived and worked here . . . and now the town is empty."

It must have been more problematical evacuating the small villages in the 30-kilometre zone. A typical example would have been Chamkov in the Gomel region of Byelorussia, which is only 6 kilometres from unit No. 4, on the opposite bank of the River Pripyat. It contains only fifty-five houses.* In all, the total population of the evacuated Byelorussian villages was 13,200. Only a very few of the evacuated villagers have been able to return to their native villages, and this occurred only after extensive monitoring both by air reconnaissance and radiation measurements and decontamination work on the ground. Two such villages were mentioned in a TASS

* In one of these lived a 16-year-old schoolgirl, Natasha Timofeyeva, who gave one of the few eyewitness accounts of the explosion. It was quite dark, and Natasha and her relatives were returning after a late visit to friends, when she saw "a bright flash over the fourth and most distant chimney of the power plant".

report of 17 July as Cheremoshnya and Nevetskoye, with the former described as "a small village with only 66 homesteads". However, by 30 December, twelve villages in the Braginsky District, Gomel Region, Byelorussia, had been repopulated, allowing some 1500 people to return. On 16 January TASS reported that it was assumed that all inhabitants from the 10-kilometre to 30-kilometre zone boundaries would also be able to return during 1987.* However, because of decontamination problems within the 10-kilometre zone, no provision has yet been made for repopulation in this area of some 300 square kilometres. It is thus entirely possible that most of the evacuated population will never return to their former homes, except perhaps to collect decontaminated property at a much later date. They will be resettled in the new towns and villages to be built specially for the evacuees.

Some of these new towns and villages will be built in the Gomel region of Byelorussia, including one in the northern part of this region. Planned for completion before October 1986, the first construction stage will contain 4000 houses and the second stage will include: schools, laundries, hospitals, kindergartens, canteens, clubs, trade centres, post offices (many postmen were evacuated with the local population and have retained their old jobs in their new environment) and other necessary facilities. Construction of this town is being carried on 24 hours a day in two shifts, and 3000 workers are employed in the task. Other localities for the evacuees are in the Kiev† and Zhitomir regions, where it is planned to build 7250 new homes and 200 consumer service and cultural projects by October 1986. In addition, 6000 privately-owned houses will be repaired to accommodate some of the Chernobyl evacuees.

It is not only new towns which have to be built. Sites also have to be prepared for the farming community evacuees. A *Pravda* report of 6 August gives some details. "Formal handing over of the keys of 150 houses for the Chervone Polissya Collective Farm Workers who moved to Modvinovka in the Makarov district, Kiev region. In each house there were placed by the builders two beds and a cot and food gifts of two bags of potatoes, cereals, jars of pickled cucumbers, and tomatoes. Each household was allotted ten hens and a pigeon loft was placed near one house." In the same district, a TASS report of 15 August told of 500 farming families having their house-warming parties and described their new accommodation and facilities as "a three- to four-roomed house, barns for cattle and poultry, plot of land for a vegetable garden and orchard, furniture, kitchen utensils: all purchased for the evacuees" (apparently, one of the "perks" of senior scientific staff at the Obninsk nuclear facility seems to be a house with a plot of garden, usually 10 metres square—gardening sometimes seems to interest them as much as science when at international conferences!).

Maternity arrangements also had to be made for the expectant mothers among the evacuees. Already, in the 30 May issue of *Izvestia*, it had been announced that forty babies had been born to evacuated mothers in the Makarovskii Maternity Home. A photograph showed seven of the babies all well wrapped up from head to toe, like Russian dolls, with only their little faces peeping out! Later, in the 30 September edition of *Pravda*, it was announced that the hundredth child had been born to "the

---

* This had not occurred by 2 December 1987.
† Fifty-two new townships were built in the Kiev region in the short period of 6–8 weeks.

large group of expectant mothers evacuated from Chernobyl", and was now living at the Ukraina Sanatorium, in a pine forest at the resort of Vorzel. The baby boy had a huge birthday cake shared by all 250 residents!

It is to be hoped that all these babies will receive medical follow-ups for the rest of their lives. This will ensure, not only that they will promptly receive any necessary treatment, but it will also provide extremely valuable data for the future on the effects of low-level radiation doses in a large population which was unborn at the time of exposure. Ideally, there should also be a comparative group of babies followed up from a region in the vicinity of a nuclear power plant, but far away from Chernobyl. These would act as an "external control group". However, at the 25–29 August 1986 IAEA meeting it appeared that the Soviet delegation was not prepared to accept the need for such a control group. Monitoring this second group throughout their lifetime would certainly have its organisational problems, but it is an opportunity which should not be missed. However, it probably will be, and future statistical analysts will have to make do with some form of "internal control group" from among the "Chernobyl babies" themselves.

There are many stories of existing villages housing evacuees in May/June 1986 before the new towns/villages were built. One was from the village of Zagaltsy which contains 800 homes. It accommodated 1000 evacuees, and although the village Soviet opened a new canteen, it was hardly used because the evacuees were fed by their hosts. One old lady of 83, Agrippina Markovets, invited several other old ladies to live with her.

During the 16–19 December 1986 visit[5] to Chernobyl by the British Government Energy Secretary, the towns of Pripyat and Chernobyl were still deserted and Mr Peter Walker was quoted in the *Guardian* of 18 December as saying that "It was a pretty depressing experience, flying over these communities, that were obviously totally empty of people. There was no activity going on whatsoever."

Fig. 63. One of the oldest evacuees, 88-year-old B. O. Vasilenko from Stechanka, 30 May 1986.

Courtesy of Izvestia

Fig. 64. Children evacuated to summer camp from the town of Pripyat.

Courtesy of TASS

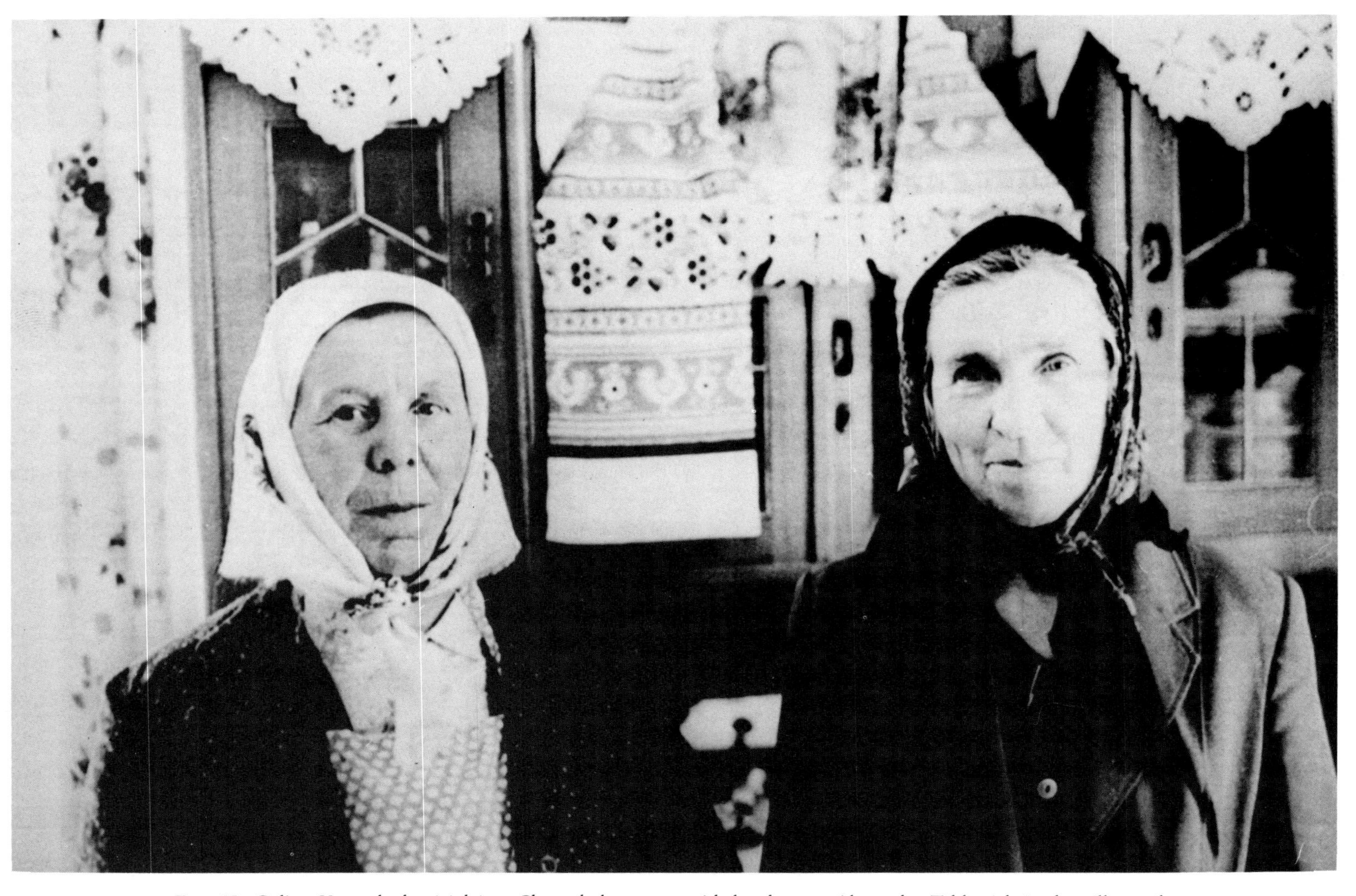

Fig. 65. Galina Yarmolenko (*right*), a Chernobyl evacuee with her hostess Alexandra Tshkevich in the village of Kopelovo, 9 May 1986.

Courtesy of Associated Press

Fig. 66. Radiation monitoring of fields in the 30-kilometre zone, May 1986.

Courtesy of Frank Spooner Agency

FIG. 67. Radiation monitoring of the railway line at the power plant, May 1986. The railway system around Chernobyl is the one transport system which received no mention in the emergency arrangements. Presumably this omission was because of very high radioactive contamination of rail facilities. For evacuation, buses and lorries were used with materials for decontamination and entombment being transported by road, helicopter and by river barge. Two male railway workers, ages 68 and 70, developed widespread beta radiation burns after standing and watching the accident from a distance of 800–1000 metres. However, after minimal medical treatment the skin burns healed.

Courtesy of Frank Spooner Agency

Fig. 68. Radiation monitoring checkpoint at the boundary of the 30-kilometre zone, May 1986. The cyrillic lettering on the sign states "Forbidden Zone".

Courtesy of TASS

Fig. 69. The commander of a decontamination unit, Major V. Tsendrovsky, May 1986. The design of the face mask used by the decontamination teams is clearly seen in this photograph.

Courtesy of Frank Spooner Agency

FIG. 70. Decontamination of houses at Chernobyl with a special solution.
Courtesy of Frank Spooner Agency

FIG. 71. Treating Chernobyl buildings with a decontaminating solution.
Courtesy of Frank Spooner Agency

FIG. 72. Decontamination of apartment blocks which once housed power plant workers, Pripyat, May 1986.
Courtesy of Camera Press

FIG. 73. There are many forests, parks and orchards within the 30-kilometre zone, and radiation level maps had to be drawn for the decontamination teams because the radioactive contamination was not uniform.
Courtesy of Novosti

FIG. 74. View of some of the forest areas close to the Chernobyl nuclear power plant.
Courtesy of Novosti

FIG. 75. Aerial view of the region near the nuclear power plant, May 1986.
Courtesy of Novosti

FIG. 76. Oleg Veklenko, a member of the Union of Artists of the USSR, shows a portrait of one of the soldiers who distinguished himself in the aftermath of the accident.

Courtesy of Pravda

FIG. 77. Soviet troops who participated in the clean-up operations, May 1986. A radiation monitor is being carried in the right hand and the associated electronics with the meters and dials to record radiation exposure are in the instrument box hung around the soldier's neck. It is noticeable that he is wearing overshoes to prevent his own boot/shoes from becoming contaminated and so presumably, at this point in time, the ground was still considered to be highly radioactive.

Courtesy of Novosti

FIG. 78. An armoured vehicle being hosed down at a decontamination centre, May 1986.

Courtesy of Camera Press

FIG. 79. Preparation of decontaminating solution used for people, clothes and equipment, 9 May 1986.

Courtesy of Associated Press

FIG. 80. A truck washes the streets of Kiev to prevent the accumulation of radioactive dust, 9 May 1986.

Courtesy of Popperphoto

FIG. 81. Decontamination of forest areas and other vegetation, May 1986.

Courtesy of TASS

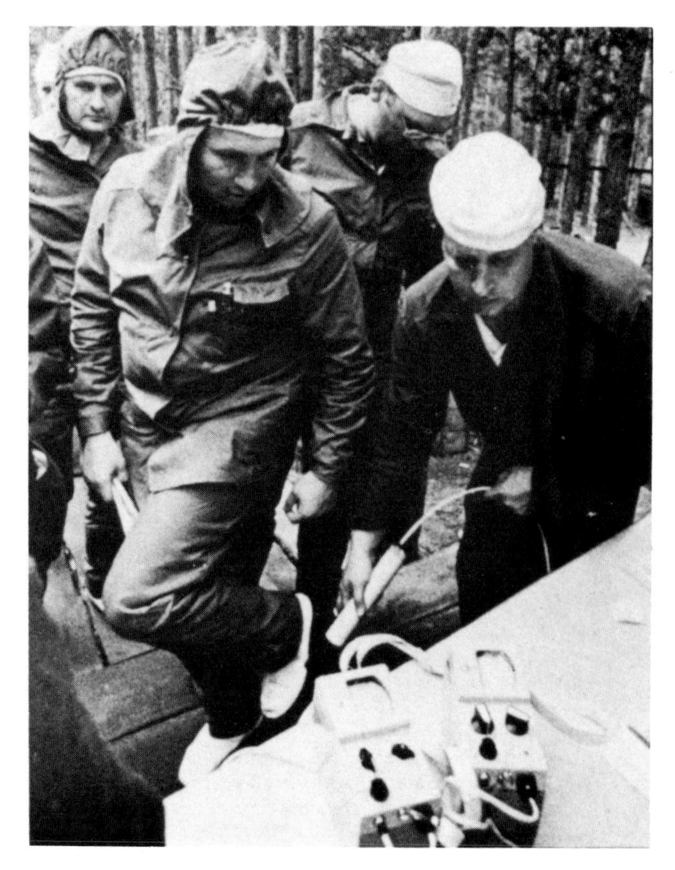

FIG. 82. Monitoring footwear of decontamination workers at Chernobyl, May 1986.

Courtesy of IAEA

FIG. 83. Decontamination of the railway lines and surroundings at the power plant, May 1986.

Courtesy of Frank Spooner Agency

FIG. 84. Remote controlled bulldozer used for earth moving, June 1986. Some of the bright yellow painted robots were described in the 19 May 1986 edition of *Pravda*. One was a giant 19-ton bulldozer with the CTZ (Chelyabinsk Tractor Works) trademark. The operator of this tractor sees it through a narrow slit in an armoured vehicle, standing several dozen metres away. It was delivered from Chelyabinsk to Kiev on board an Ilyushin-76 jumbo jet, and specialists from institutes in Kiev outfitted it with sophisticated electronic equipment. In *Pravda*, 10 October, it was reported that in the territory of the plant 30 centimetres of soil was removed and the area covered with hermetically sealed concrete slabs. Also, that it had taken 2 months to decontaminate the power plant site.

Courtesy of TASS

FIG. 85. Remote controlled earth-moving machine, May 1986.

Courtesy of TASS

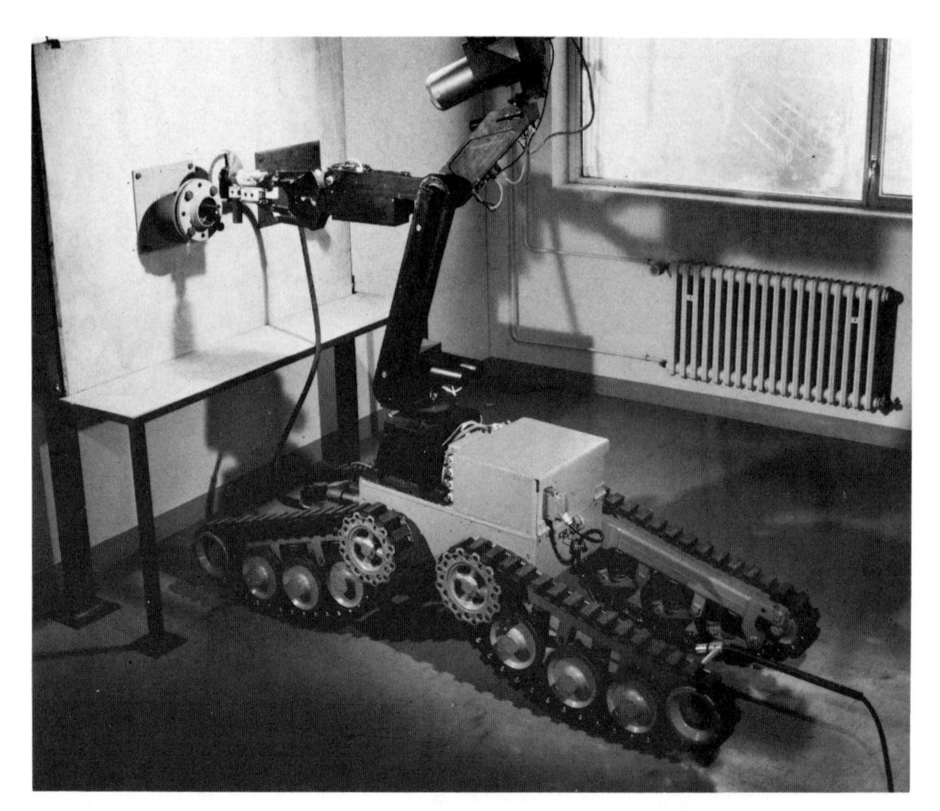

FIG. 86. Three types of remote controlled robot were sent to Chernobyl by the Kerntechnische Hilfsdienst GmbH in the Federal Republic of Germany. This is the MF3 miniature robot with a height of only 0.4 metre and a width of 0.745 metre, but with seven degrees of freedom of movement for the manipulator arm. It is a cable-controlled tracked vehicle and carries a folding arm manipulator, a black and white stereo TV camera and additional cameras, microphones and lighting equipment. Tongs are also available which can accommodate a number of tools. The multi-core control cable is 100 metres long.

Courtesy of Kerntechnische Hilfdienst GmbH

FIG. 87. A Kerntechnische Hilfdienst radio-controlled bucket loader, 6.7 metres long, 2.4 metres wide and 2.75 metres high. It is a wheel loader with articulated steering and is driven by a 125 horsepower diesel engine coupled to a ZF Hydromedia transmission with a torque converter and a reversing gear. Another diesel engine of 70 horsepower drives a double-vane cell pump supplying the hydraulic working system.

Courtesy of Kerntechnische Hilfdienst GmbH

FIG. 88. One of the Kerntechnische Hilfdienst robots used at Chernobyl.

Courtesy of Kerntechnische Hilfdienst GmbH

FIG. 89. What appears to be a view of the removal of contaminated earth, May 1986.

Courtesy of Frank Spooner Agency

Fig. 90. Colonel-General Vladimir Pikalov, Commander of Chemical Forces, USSR Ministry of Defence, from a photograph published in the 25 December 1986 issue of *Pravda*, under the heading "Heroes of Chernobyl". General Pikalov later wrote his own story, "Harsh Lessons of Chernobyl Disaster", in a Special Issue of *Military Bulletin* (No. 8, Vol. 14, April 1987), reprinted in English by the Novosti Press Agency. He summarises the decontamination achievements of the Chemical Forces one year after the accident in the following statistics: "More than 500 residential communities, nearly 60,000 buildings and structures and several dozen million square metres of exposed surfaces of technological equipment and internal surfaces at the [nuclear power] station itself have been decontaminated. Tens of thousands of cubic metres of contaminated soil have been removed and the same amount brought in, and several thousand insulating screens have been laid down. Dust has been suppressed on vast territories and several thousand samples have been taken for [radioactive] isotope analysis." Soil contamination is described as follows: "Today [April 1987], higher than admissible readings of soil contamination with long-living radionuclides [caesium, strontium, plutonium] are registered mostly on the territory of the station and in the 5-kilometre zone around it, as well as in several pockets on the territory of Byelorussia. There are no grounds for expecting any radical changes in the current radiation situation, since soil radioactivity flushing by flood waters has not exceeded 1%." General Pikalov also refers to nuclear warfare and specifically comments on "the explosion of a nuclear weapon hitting a nuclear power station" and uses the words "in the case of nuclear war in Europe, which will certainly cause the destruction of most atomic power stations and of nuclear processing plants". When discussing this scenario (which I have not often seen mentioned), he puts the accident into perspective in terms of fallout by stating that "the Chernobyl disaster is just a scaled-down version of a low yield nuclear explosion involving only radioactive fallout without the devastating shock wave, thermal or penetrating radiation".

Courtesy of Pravda

Fig. 91. Also from the 25 December 1986 issue of *Pravda* (see Fig. 90) was this photograph of two members of the Chemical Defence Force, Junior Sergeant A. Murtazaliev and Major S. Shishko. Vehicles like the one in this photograph must have been used during the clean-up after the accident with a certain amount of protection afforded by the tank-like turret shielding.

Courtesy of Pravda

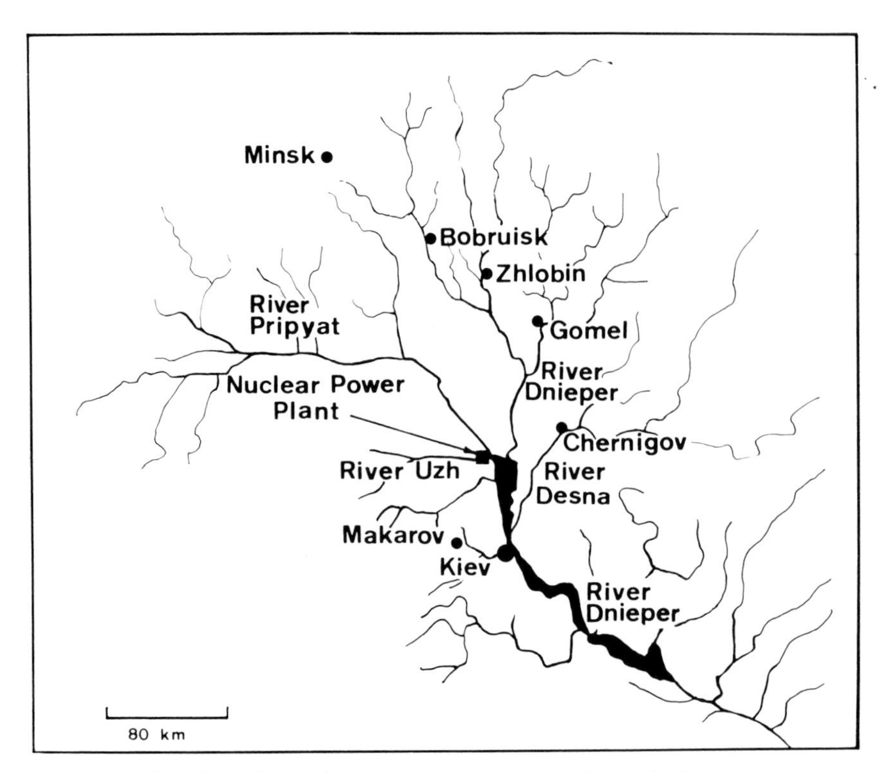

FIG. 92. The Chernobyl nuclear power plant, towns and cities in the Ukraine and Byelorussia outside the 30-kilometre evacuation zone, and the river system of the Dnieper, Pripyat, Desna and Uzh, which are the major rivers in this region.

FIG. 93. A view of the Dnieper river from Vladimir Hill, Kiev.

Courtesy of Novosti

FIG. 94. Drillers help themselves to the first mouthfuls of water out of an artesian well they have just sunk in Kiev, 1986.

Courtesy of Novosti

FIG. 95. The pump ship *Rosa-300*, moored in the Desna to provide clean water for Kiev. Two emergency water supply mains, each 6 kilometres long, were built within a month to provide water for Kiev in the event of radioactive contamination of normal supplies. Both mains are fed by the water of the River Desna, a tributary of the Dnieper. Pipelines to the city had to be thrown over 18 obstacles including rivers, bridges, tunnels and roads.

Courtesy of Novosti

FIG. 96. *Courtesy of Associated Press*

FIG. 96. Army personnel by the River Pripyat after a successful completion of a pontoon bridge over the river. Similar pictures appeared in *Pravda*, 15 May 1986. Later, in *Pravda*, 25 December 1986, the initial response and contribution by the military was described by a *Pravda* correspondent under the headline "Zone of Responsibility: The Chernobyl Heroes", as follows: "At the end of April 1986, Colonel-General Vladimir Pikalov, chief of the USSR Defence Ministry's chemical defence troops, was far away from Moscow to attend a scheduled combat training exercise. Before midnight of 25–26 April, the General and his subordinates set off for a rest. They did not yet know, and who could have foreseen it, that at 3.12 am the alarm signal would raise, among others, the chemical units of the Kiev and some other military districts. The assignment seemed to be totally unexpected and extraordinary. The first combined detachment of chemical defence troops rushed to Chernobyl. The General Staff summonded the chief of the chemical troops in the morning. At 2 pm General Pikalov and members of his chemical task force climbed down the ramp of an AN-26 transport onto the concrete of Kiev's Zhulany Airport. Meanwhile, several heavy transports already were flying reinforcements to the Kievans, the advanced chemical group led by Lieutenant-Colonel N. Vybodovsky. The mobile detachment's main forces under Major V. Skachkov were entraining. On 2 May, Party Politburo members arrived at the Chernobyl Nuclear Power Plant. After reports from specialists, including Pikalov, they decided on a strategy to quell the effects of the accident. Army units were among the first on the scene. The officers and men helped evacuate the population, conducted radiation reconnaissance, started complex engineering work and prepared for large-scale decontamination. The most important thing was to master the situation, especially near power unit No. 4 and the adjacent territory, literally on each square metre, to suggest reactor seal-off variants and to ensure population safety. Chemical units were still being deployed in assigned areas when Senior Lieutenants V. Blokhin and A. Toporkov had already led their platoons to the plant site for radiation reconnaissance. By the end of 26 April, General Pikalov already had his own 'corner' at the Pripyat city party committee, with a table and an intercom (later his group moved to Chernobyl). Information started pouring in from radiation monitoring posts. No one had a wink of sleep that night. At 8 am on 27 April representatives of chemical defence troops, together with nuclear specialists, were already reporting the situation to members of the Government Commission. By then it had sharply deteriorated. What happened next is known now."

The *Pravda* article also described General Pikalov's first reconnoitring of the site on the night of 27 April: "I was in a Volga, for it was essential to get in the situation at once. I understood that even if there were a radioactive discharge at that moment, we would get through fast. The fire had already been put out and there was dead silence around. The air above the 4th unit was aglow and in principle, the cause of that devilish glow was clear. So what was to be done next? In an armoured vehicle I rode to the reactor. At a safe distance I ordered the driver out and took the wheel myself. I drove up to the gates and smashed them in. The vehicle stuffed with sensitive instruments stopped at the entrance to the ruined reactor."

FIG. 97. Monitoring fish from the River Dnieper for radioactive contamination, 8 June 1986.

Courtesy of Pravda

FIG. 98. A researcher from the Kurchatov Atomic Energy Institute surveys the damaged unit No. 4 during the clean-up and entombment work, October 1986.
Courtesy of TASS

FIG. 99. A helicopter hovering over the roofs of units Nos. 3 and 4 with dust-suppression equipment, October 1986.
Courtesy of TASS

FIG. 100. An aerial view of the damaged unit No. 4, September 1986.
Courtesy of TASS

FIG. 101. Environmental monitoring.

Courtesy of TASS

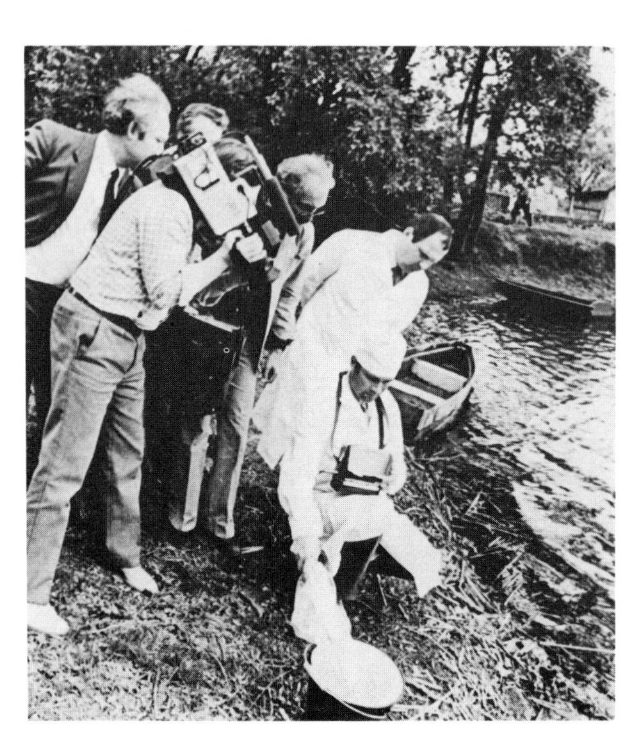

FIG. 102. View of the power plant, October 1986.
Courtesy of TASS

FIG. 103. The first TASS photograph showing inside the damaged building. It was taken during the construction of the protective barrier separating units Nos. 3 and 4, September 1986. The scale of the building can be envisaged by comparing its height with the man at bottom-centre of the photograph.

Courtesy of TASS

FIG. 104. The damaged unit No. 4, September 1986.

Courtesy of TASS

FIG. 105. Firemen from Chernobyl talking with doctors at the rehabilitation centre outside Moscow, which is part of Moscow's Central Hospital. The caps cover bald scalps which have been shaved as part of the decontamination process.

Courtesy of TASS

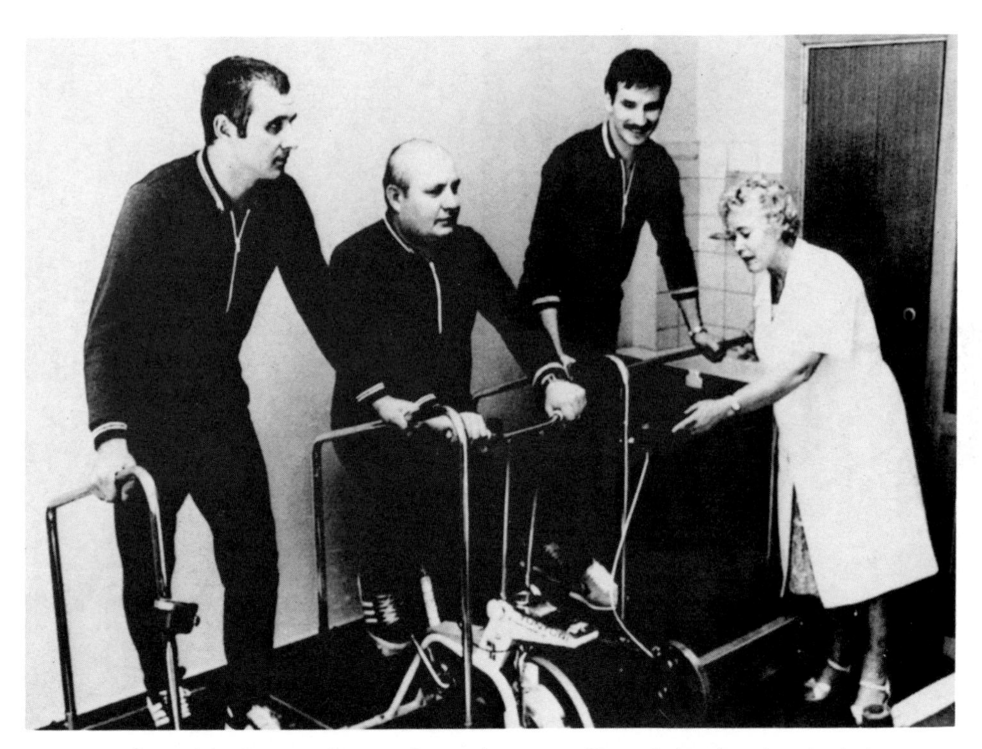

FIG. 106. Three of the firemen who were first on the scene at Chernobyl undergoing physiotherapy at a hospital of the USSR Internal Affairs Ministry near Moscow. *Left to right* are Junior Sergeant R. Polovinkin, Senior Sergeant V. Prishchepa and Ensign V. Palagecha, 5 July 1986. The central piece of equipment is a Puch Tunturi static exercise bicycle and the other two pieces appear to be treadmills for measurements of cardiac capacity or of lung capacity. Lung capacity of the firemen could well have been impaired due to inhalation of smoke, dust or dirt.

Courtesy of Associated Press

FIG. 107. Cement mixers on their way to Chernobyl. One of the local houses is clearly seen on the left, May 1986. Eventually, some 400,000 tonnes of concrete were to be used for the entombment of the reactor.

Courtesy of TASS

FIG. 108. Site assembly of the metalworks for the protective wall, September 1986. This was 10 kilometres from the power station and remains of some of the scaffolding were still to be seen on 2 December 1987.

Courtesy of TASS

FIG. 109. Assembly of the metalworks for the protective wall around unit No. 4, September 1986. The final sarcophagus was to contain more than 6000 tonnes of metal structures.

Courtesy of TASS

FIG. 110. An aerial view of the power plant, September 1986.
Courtesy of TASS

FIG. 111. A close-up view of unit No. 4, September 1986.
Courtesy of TASS

FIG. 112. A close-up view of unit No. 4, October 1986.
Courtesy of TASS

FIG. 113. Construction of the protective concrete wall around unit No. 4, October 1986. This clearly shows the extent of the damage.

Courtesy of TASS

FIG. 114. Deputy B. Knurenko holds up the symbolic key to the settlement of Zelyony Mys which was built to house the clean-up and service personnel following the accident.

Courtesy of Novosti

FIG. 115. Housing for evacuees in the village of Pologiya Verguny in the Cgerkassy region of the Ukraine, August 1986.

Courtesy of TASS

6

Decontamination of the Environment

"The Russians proved to be dedicated, gregarious, competent people who handled the situation extremely well."
(Morris Rosen, Director of the Division of Nuclear Safety, IAEA, at the 54th Annual Edison Electric Institute Convention, San Francisco, USA, June 1986, after visiting Chernobyl with Dr Hans Blix)

THE SOUNDTRACK of the Soviet television videotape which was quoted at the start of the previous chapter also contained the following passages which describe some of the problems involved with the decontamination procedures around the power plant, as well as further afield within the 30-kilometre evacuation zone:

"The work is organised on a rotation basis; people live and work in the 30-kilometre zone for 15 days and nights. This is followed by 15 days of rest with special medical check-ups, and then back to the station. Working shifts vary from 8–10 hours down to a few minutes, depending on the level of radiation. Armoured personnel carriers are used to transport people to especially hazardous areas. More and more remotely-controlled vehicles are arriving here to work in high radiation areas. The hardware is coming from all over the country; a powerful force is massed here to wipe out the aftermath of the accident as soon as possible. Special roads have been made in the zone and all the vehicles keep within their limits to prevent the spread of radioactive matter. They are treated at special decontamination centres. Work is also underway to decontaminate the 30-kilometre zone. It is divided into three sectors. Data comes several times a day from 240 points on the ground and in the air. It is now known that the radioactive contamination represents a chequered pattern. Radiation monitoring is being done on an around-the-clock basis inside the buildings housing the power units and in the plant area. Embankments have been set up around the plant's area. A protective wall along the river bank prevents radioactive substances reaching the river in torrential rain. The subterranean waters are being placed under full control. Automatic systems are being set up to cleanse, dump, and distribute subterranean water. Here a protective wall is being erected in the ground to prevent the outflow of underground water from the plant area. The wall stands 30 metres high and [there is] more underground. Helicopters are decontaminating the buildings of the Chernobyl plant and surrounding terrain. Specially trained soldiers are decontaminating homes and other buildings, making wide use of liquid polymase. Decontamination is gruelling work, requiring much patience, as the wind moves dust from a contaminated area and the work has to be done all over again. Robots are used in areas with a high level of radiation. They look for and remove radioactive fragments with the help of television cameras and dosimeters. No matter how long a person may be in the zone, it [radioactive radiations] can conjure up a danger without taste, colour or smell; it can only be detected by special instruments. All the people who have returned to the 30-kilometre zone have undergone rigorous medical control."

The problems of decontamination in any short-term period were insurmountable when one considers that this related not only to the nuclear power plant site—which had to be made radiation hazard-free if any of the remaining units, particularly No. 3, were ever to be put into operation again for electricity production—but also to the 30-kilometre zone. The latter included not only buildings and transport, but roads, soil, forests and other vegetation. In addition, wind and rainfall paterns could obstruct the decontamination task, and if any rivers and lakes became contaminated there would be the necessity of stopping the contamination from spreading further along waterways. The problem of where to bury all the radioactive waste had also to be solved, as it was in part by using the pit excavated for unit No. 5.

The nuclear power plant site was contaminated over a wide area, and radioactive material was scattered on to the roof of the turbine hall, the roof of unit No. 3 and on metal pipe supports—of which there were many. The site as a whole, as well as walls and tops of buildings, was substantially contaminated from radioactive aerosols and radioactive dust, but the contamination was not uniform. With a view to reducing the spread of radioactive dust, the roof of the turbine hall and the road shoulders were treated with various polymerising solutions to immobilise the upper layers of the soil. This prevented any dust from rising.

The ventilation system continued to operate for some time after the accident to unit No. 4, and as a result, radioactive dust was spread over the surfaces of the equipment and the compartments of the nuclear power plant. The highest levels were recorded for horizontal sections of surfaces in the turbine hall, since contamination was caused here not only via the ventilation system but also because material could fall through the damaged roof. On 20 May the gamma-radiation exposure rate in the contaminated compartments of units Nos. 1 and 2 was 10–100 millirontgens/hour and in the turbine building was 20–600 millirontgens/hour. After decontamination of units Nos. 1 and 2, the exposure rate in the compartments fell to a level of 2–10 millirontgens/hour. A decontamination spraying method was widely used for the compartments, employing washing machines and fire hydrants, although some compartments had to be washed manually using rags soaked in decontaminating solution. The composition of these were chosen bearing in mind the contaminated surface, which might be plastic, steel, concrete or various other coverings.

Decontamination of the plant area was undertaken in the following sequence:

1. Removal of refuse and contaminated equipment from the site.

2. Decontamination of roofs and outer surfaces of buildings—sometimes, pastes were put on walls so that they established a quick-drying film, which when peeled off had radioactive particles sticking to it.
3. Removal of a 5–10 centimetre layer of soil and its transfer in containers to the waste disposal dump at unit No. 5 site.
4. Laying, if necessary, of concrete slabs on the soil or filling with clean earth.
5. Coating of the slabs and of the non-concrete areas with film-forming compounds.

As a result of these measures it was possible (USSR Delegation, 25–29 August 1986) to reduce the total gamma-radiation background in the area of unit No. 1 to an exposure rate of 20–30 millirontgens/hour. This was due mainly to the damaged unit No. 4. The work rate of the decontamination teams was such that the surfaces were cleaned at a rate of 15,000–35,000 square metres per 24 hours.

In the 30-kilometre evacuation zone it was decided to establish three surveillance zones: a special zone, a 10-kilometre zone and a 30-kilometre zone. Strict dosimetric monitoring of all transport was organised and decontamination points were established by the militia. At the zone boundaries, arrangements were made to transfer working personnel from one vehicle to another in order to reduce the possibility of transferring contamination across zones. A Novosti report in July from the city of Zhlobin, some 180 kilometres from Chernobyl, gives some idea of traffic build-up. The correspondent was travelling to Minsk, a journey which normally took 2½ hours. "It took somewhat longer, and past Bobruisk, which was 80 kilometres from Zhlobin, a line of lorries, refrigerator cars and Ladas caught our eye. They had been stopped by traffic militia for radiation check-up. The militia captain said that they thoroughly hose down vehicles coming from the south when the radiation exposure is over 0.3 millirontgens/hour. In the first days following the accident, he had to hose down nearly every car. Today (July) it has only been thirty and the overall radiation level is 0.025 millirontgens/hour." Zhlobin is the site of Byelorussia's first steel works, which now annually outputs some 500,000 tons of steel.

The radiation contamination within the 30-kilometre zone will continue to change, due to wind and rainfall patterns, and a substantial redistribution of radioactivity over the landscape may occur. In the case of coniferous forests, redistribution of radioactivity can be expected to end only after some 3–4 years when complete renewal of the pine needles has taken place. One of the fears concerns forest fires, and because of this increased precautions are being taken in forest areas to prevent fire which would redistribute radioactivity on leaves and forest litter. In a report in *The Times* of 2 July 1986, based on an item in *Komsomolskaya Pravda*, it was stated that "Peat fires hit areas north of Chernobyl" and described a fire in the Gomel region of Byelorussia, along a 1-mile front, where the situation was such that the firemen needed special breathing apparatus.

Examples of some of the problems encountered in this major programme of decontamination were:

Localisation of the radioactive contamination, especially because of the very large number of vehicles necessarily involved in clean-up operations.

Loose caesium-137 contamination.

> *Disposal of contaminated clothes and provision of new clean clothes, especially since at one stage near the plant there were 1000 persons in protective clothing, along with all sorts of equipment, including many concrete mixers.*
>
> *Although thousands of square metres were sprayed daily with inexpensive non-toxic substances, wind erosion of firstly roads then soil and crops was substantial.*

Construction of a complex of hydraulic engineering structures began with a view to protecting the ground water and surface water from contamination. This included a filtration-proof wall in the soil along part of the perimeter of the industrial site of the power plant, and wells for lowering the water table; a drainage barrier in the cooling pond; a drainage cut-off barrier on the right-hand bank of the River Pripyat; a drainage interception barrier in the south-western sector of the power plant; and drainage water purification facilities.

Geographical details were given by TASS, 31 October 1986, mentioning the rivers and quoting lengths of the dams. "In the Kiev reservoir a high-powered dredger is building an underwater bar 450 metres long across the fairway, in front of it. There will be a hollow, 100 metres wide and about 16 metres deep, that will be an underwater trap, the purpose of which will be to catch radioactive isotopes which may get through from the Pripyat river tributaries. Another silt trap has been made in front of the dam of the Kiev hydropower station. Yet another underwater ridge has been built from stone to the north of the mouth of the River Pripyat, in the area of Atashev. The construction of the water protection facilities has been on the Rivers Sakhan, Ilya, Veresnya, Berezhest, Radynka, Braginka, Nesvich and other small rivers which fall into the Uzh and Pripyat. Such facilities have also been built on soil reclamation canals. The overall length of all structures, dams and coffer dams, is 29 kilometres of earth and sand. Almost a quarter of a million cubic metres of rock mass has been placed into water. A large amount of work has also been undertaken on drainage canals. The purpose of each of these facilities is one and the same, to prevent contamination of water flowing from the Pripyat to the Dnieper, during autumn and spring flood time."

As soon as possible after the accident, two aeroplanes, with helicopters and motorised vehicles, were deployed for monitoring measurements over 20,000 square kilometres, and an enormous number of samples were taken for analysis from soil, waterways and air. An aerogeologia (aerial geology) unit was formed from an amalgamation of departments from the Kurchatov Atomic Energy Institute and the Terrestrial Physics Institute of the USSR Academy of Sciences, and by 2 September (TASS, *Izvestia*) an area 110 square kilometres, centred on Pripyat, had received an aerial survey. The base of operations was Chernigov, and the instruments were "tuned" above the Kiev sea. Flying was in a south-to-north direction at an altitude of 170 metres, with a speed of 150 kilometres per hour. The operation was completed between 5 July and 20 August, when a report was submitted to the government commission investigating the accident. *Izvestia* reported that a further 10,000 square kilometres were to be surveyed in September. The aerial photography helped to establish suspect areas of contamination, and landing could then take place and samples taken for analysis.

Little data on food contamination or recommended precautions to be taken with food were given by the Soviet delegation during the 25–29 August 1986 meeting, or indeed submitted earlier to the World Health Organisation for its survey published after 12 June (see Chapter 7). The following details of instructions concerning food were given by *Pravda* on 20 May: "People such as anglers and those who like to gather mushrooms and berries are warned that the 30-kilometre zone has been sealed off and they should not try to penetrate it." Then on 2 May, *Izvestia* interviewed the Head of the Central Sanitary and Epidemiologic Service of the Ukraine, had this to say: "Departing passengers are subject to thorough dosimetric inspection at airports, railway stations, and long distance bus stations. Food is subject to strict control and temporary severe requirements as to quality, throughout the Ukraine. Thousands of ice cream, cake and soft drink stalls have vanished from the streets of Kiev and food is no longer sold in the open air."

The disappearance of the Russian's favourite ice cream stalls apparently caused some panic (as well it might) and was reported by *The Times* of 9 May: "Kiev residents say that the unease began to gather momentum at the beginning of the week when the local health authorities broke an earlier silence and issued televised warnings against letting children outside for more than short periods and against eating leaf vegetables. The mood grew more anxious after a change in the direction of the wind resulted in new orders to wash down all interiors of flats with cold water and to avoid swimming in reservoirs. The panic had been increased by new measures to impose radiation checks on all those leaving and by a ban on street sales of ice cream, cakes and drinks reported by *Izvestia*."

Panic was also reported on trains leaving Kiev, and doubtless there certainly was some panic, but all newspapers cannot be correct, as there is some difference between 2000 and 250,000, as is shown in what follows.

Daily Mail newspaper in the United Kingdom, 8 May

"Flight from Kiev: more than 2000 children arrived in Moscow within minutes yesterday as three trains pulled in from the frightened city of Kiev. . . . Children have been coming into Moscow every day since Saturday" (8 May was a Thursday).

London Evening Standard, 9 May

"250,000 children flee Kiev fall-out."

The Times, 9 May

"Trains packed with fleeing families" and ". . . the overnight express from Kiev[6] pulled in at (Moscow) 9.45 am yesterday . . . with hundreds of women and children spilling onto the platform. . . ."

The second mention of food recommendations by *Izvestia* was on 9 June under the headline "The Dnieper Clean". "People can swim and use beaches, but basic rules are optimal time in the sun, 8 am to 11 am, and one should not stay in the sun too long; wear a hat or cap when on the beach or stay under a sunshade; also use trestle beds, chaise longues and rugs which can be found on the beach; more important, it is not recommended to play volleyball or football, because of the dust that is kicked up during play; eat on the beaches only in specially fitted out places; observe the rules of hygiene." *A Moscow News* issue in September stated that "large quantities of contaminated vegetables and fruit have had to be destroyed in the Ukraine and in Byelorussia."

The famous Russian firewater is vodka (although each Soviet republic also has its own brand of cognac, and there are fine Georgian wines and even red champagne from Azerbaijan), and reports were given in the press* of it being a potent anti-radiation folk medicine if taken with strong red wine. A bottle of real Kiev vodka, called Kievskaya Gorilka, is shown in Fig. 41. It was presented to me in Geneva in July 1986 with the comment that it was made with the water from Chernobyl. The label translates to say that the bottle contains special firewater made from the ancient recipes, and includes berries and honey, and is best taken ice cold. The bottle is still intact!

The Soviet estimate of fission product (excluding radioactive noble gases such as xenon and krypton) radioactive release was $3\frac{1}{2}$% of the radioactive core material (that is, radioactive isotopes within the reactor) at the time of the accident. This amounted to a total activity of approximately 50 megacuries and was some 6–7 tonnes of material. A list of the radioactive isotopes is given below.

Assessment of Radioactive Isotope Composition of the Release from the Damaged Unit No. 4 (accuracy of assessment is 50%)

Isotope	Release	Half-life	Decay mode
Xenon-133	? Up to 100%	5.3 days	beta+gamma
Krypton-85m	? Up to 100%	4.4 hours	beta+gamma
Krypton-85	? Up to 100%	10.7 days	beta
Iodine-131	20%	8.0 days	beta+gamma
Tellurium-132	15%	3.3 days	beta+gamma
Caesium-134	10%	2.1 years	beta+gamma
Caesium-137	13%	30.1 years	beta+gamma
Molybdenum-99	2.3%	2.8 days	beta+gamma
Zirconium-95	3.2%	64.8 days	beta+gamma
Ruthenium-103	2.9%	40.0 days	beta+gamma
Ruthenium-106	2.9%	371.6 days	beta+gamma
Barium-140	5.6%	12.8 days	beta+gamma
Cerium-141	2.3%	32.5 days	beta+gamma
Cerium-144	2.8%	284.9 days	beta+gamma
Strontium-89	4.0%	50.6 days	beta
Strontium-90	4.0%	28.6 years	beta
Plutonium-238	3.0%	86 years	alpha
Plutonium-239	3.0%	24,100 years	alpha+gamma
Plutonium-240	3.0%	6560 years	alpha+gamma
Plutonium-241	3.0%	14.4 years	beta
Plutonium-242	3.0%	380,000 years	alpha+gamma
Curium-242	3.0%	163 days	alpha+gamma
Neptunium-239	3.2%	2.4 days	beta+gamma

This assessment was based on: systematic analyses of the radioactive isotope composition of aerosol samples collected at points above the damaged reactor from 26 April 1986 onwards; airborne gamma-radiation surveys of the plant area; analysis of fall-out samples; systematic measurement data from the meteorological stations of the Soviet Union.

The distribution of fall-out was estimated to be 0.6–1 tonne in the immediate area of the power plant, 3–4 tonnes out to the 20-kilometre zone, 2–3 tonnes at greater distances. Particle sizes ranged from less than 1 micron to tens of microns.

However, not everyone had agreed with the Soviet estimates, and, for example,

* *The Times*, 23 May: "Flights of alcoholic fancy in Ukraine"; *The Times*, 9 July: "Russians still use drink as the nuclear deterrent."

Collier and Davies (1986) put the release figure at 70 megacuries and not at 50 megacuries. These authors have also given an estimated comparison between the Chernobyl and Three Mile Island releases.

Comparison of Chernobyl and Three Mile Island Estimated Releases (after Collier and Davies, 1986)

Isotopes	Three Mile Island		Chernobyl
	Outside core	To the environment	To the environment
Noble gases	48%	1%	100%
Iodine	25%	0.00003%	20%
Caesium	53%	Not detected	10–15%
Ruthenium	0.5%	Not detected	2.9%
Cerium	Nil	Not detected	2.3–2.8%

The priority objectives of the environment monitoring systems were: (1) assessment of the possible internal and external exposures of plant personnel, the population of Pripyat and of the 30-kilometre evacuation zone; (2) assessment of possible exposure of the population of a number of areas outside the 30-kilometre zone, the level of radioactive contamination which could have exceeded the permissible limits; and (3) preparation of recommendations on measures to protect the population and staff from exposure in excess of the established limits. These recommendations included: evacuation of the population; restrictions or a ban on the use of food products containing increased amounts of radioactive substances; and recommendations on what action people at home and in open places should take.

The Soviet delegation papers of 25–29 August 1986 detailed what systematic radiation monitoring was undertaken to solve these priority problems, including:

The level of gamma-radiation in contaminated areas.

The concentration of biologically significant isotopes in the air and in water sources, particularly those supplying drinking water.

The degree of radioactive contamination of the soil and vegetation and its radioactive isotopic composition.

The amount of radioactive substances in food products, especially iodine-131 in milk.

The radioactive contamination of working and non-working clothes; footwear and means of transport.

The build-up of radioactive isotopes in the internal organs of people.

Some of this work will be co-ordinated and analysed by a new radiological centre which has been set up in Kiev (*Trud*, 11 September). It consists of three research units, an Institute of Experimental Radiology, an Institute of Clinical Radiology and an Institute of Epidemiology and Prevention of Radiation Sickness. A register of patient records is planned, giving the results of regular health examinations, and people in areas of increased background will be monitored. Pharmacists and doctors will be looking for new and more effective drugs to treat radiation sickness and also for drugs to prevent the accumulation of radioactive isotopes in human tissue. The centre will have a 600-bed clinic and several laboratories. A separate Novosti report in September about this Institute of Radiological Medicine stated that nearly 6000 doctors and more than 10,000 paramedics have been enlisted for a "health scanning campaign".

The radioactive plume moved first to the west and north during the 2–3 days after the accident, and, for a few days from 29 April, to the south. The contaminated air masses then dispersed for great distances over the Byelorussian, Ukrainian and Russian republics. The plume also carried radioactive contamination at a level much lower than that experienced in the Soviet Union to many European and Scandinavian countries. This prompted many of these countries to carry out contamination checks of passengers and foodstuffs from Eastern Europe and radiation monitoring of vehicles at border crossings. For example, airport monitoring in Vienna is shown in Fig. 48, as is the interior of the whole-body monitoring caravan of the British National Radiological Protection Board in Fig. 49. This caravan was sent to London's Heathrow Airport to await the return of some of the British students from Kiev and Minsk. However, these radiation checks, coupled with a general lack of availability of explanatory information, caused much concern and, for a few days, not a little panic.

On 2 December 1987, shortly after I was leaving the power station by bus for Pripyat, a single tree stood out in the snow, peculiarly shaped like a trident, and surrounded by a very low fence and close to what seemed to be a few grave stones and red wreaths. It was the only living thing in sight and had been left as a memorial – to those Ukrainians hung from its branches by the Nazis in the Great Patriotic war of 1941–1944.

7

The Food Chain

"The sensitivity of the thyroid to the induction of cancer by radiation appears to be higher than that of the red bone marrow to the development of leukaemia. However, the mortality from these thyroid cancers is much lower than for leukaemia, primarily because of the success in the treatment of thyroid cancer and the slow progress of this type of tumour.

The overall mortality risk factor is considered to be about one-quarter of that for the red bone marrow; for radiation protection purposes the risk factor is taken as 5 per 10,000 sievert."
(*Recommendations of the International Commission on Radiological Protection*, ICRP Publication 26, 1977, Clause 56.)

ENTRY INTO the food chain of radioactive material such as the relatively short-lived radioactive iodine-131, with its half-life of 8 days, or the long-lived caesium-137 (half-life 30 years), was the immediate worry for most of those who had possibly been affected, with milk, vegetables and fruit providing the greatest concern. This was particularly true of parts of Bavaria in the Federal Republic of Germany where there was coincidentally a very heavy thunderstorm on the afternoon of 30 April 1986 at the time the Chernobyl cloud was passing. The following washout contamination rate measurements of grass illustrate the radioactive deposition which occurred near Munich in a few minutes.

Prior to 30 April 1986	0.08	microsievert/hour
Afternoon of 30 April	1	microsievert/hour
At end of 1986	0.12	microsievert/hour

The mechanism of human exposure to contaminants from the Chernobyl accident have been listed by Fry and colleagues in their paper which appeared in *Nature*, 15 May, on early dose estimates in the United Kingdom. They are (1) external irradiation by the radioactive cloud; (2) inhalation of radioactive material in the cloud; (3) beta-radiation contamination of the skin; (4) external irradiation by material deposited on the ground; (5) ingestion of contaminated foods. For mechanism (5), contribution from iodine-131 will have been through milk from cows which have eaten contaminated grass or fodder and through eating leafy green vegetables which have been contaminated. Pathways of caesium-137 (or caesium-134, of which there was much less fall-out than the isotope 137) into growing plants are either by deposition on leaves or by uptake from soil through roots. In addition, with caesium-137 special consideration must be given to game meat, such as deer and rabbits, and, in particular, reindeer, where the concentration can be particularly high because of their diet of lichens. Caesium-137 activity in fish might also rise significantly in freshwater lakes, although not in seawater or in estuary fisheries.

Once these radioactive isotopes are ingested, then the iodine locates itself in the thyroid and the caesium is almost completely absorbed by the gastrointestinal tract. Strontium-90, which from Chernobyl is only some 1% of the total caesium-137 activity, is also absorbed by the gastrointestinal tract.

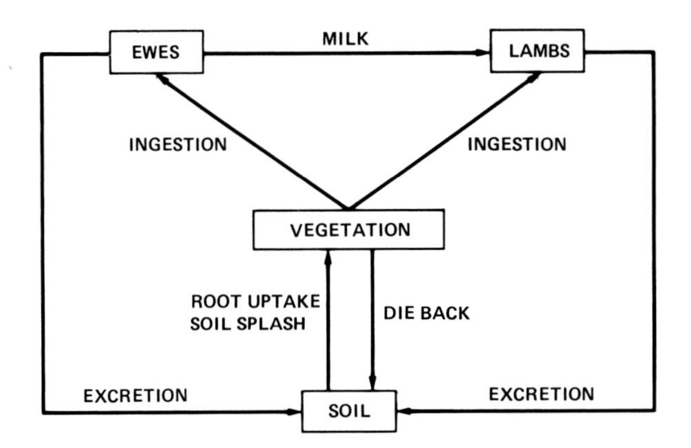

Simplified caesium cycle schematic for soil–vegetation–sheep–soil. After Dr F. R. Livens, who spoke on "The Radio-ecology of Chernobyl Fallout in Britain" at the 11 April 1987 seminar in London of the Institute of Biology. Livens also stated that in upland soil in Cumbria, the county which contains Windscale (site of the 1957 accident), the total caesium-137 activity is 470 Bq/kg, of which 155 Bq/kg is considered to be due to Chernobyl. The remainder is due to weapons testing and Windscale itself.

The contamination from the Chernobyl radioactive cloud, first noted early on 28 April at the Forsmark nuclear power station, 100 kilometres north of Stockholm, depended on wind and rainfall patterns. The heat from the reactor core would have made the radioactive plume rise, and the dry weather at the time over Pripyat was such that it was reported to have risen to a height of 1200 metres on 27 April 1986. The plume travelled towards Finland and Sweden, becoming stagnant for some time over the Ukraine and north-eastern Europe, with some rainfall over Scandinavia, and then spread widely, including passage over parts of the United Kingdom, France and other neighbouring countries. On 1 May, when the plume was over southern Germany, there was particularly heavy rainfall, resulting in substantial contamination of soil and vegetation several times higher than that measured some 20 years earlier at the time of nuclear weapons testing in the atmosphere, which ended in the mid-1960s. The reindeer problem became, due to rainfall patterns, most acute in areas of northern Sweden, where thousands of animals had to be slaughtered, each normally worth some £150 to £200. However, a large tract of the Finnish Lappland

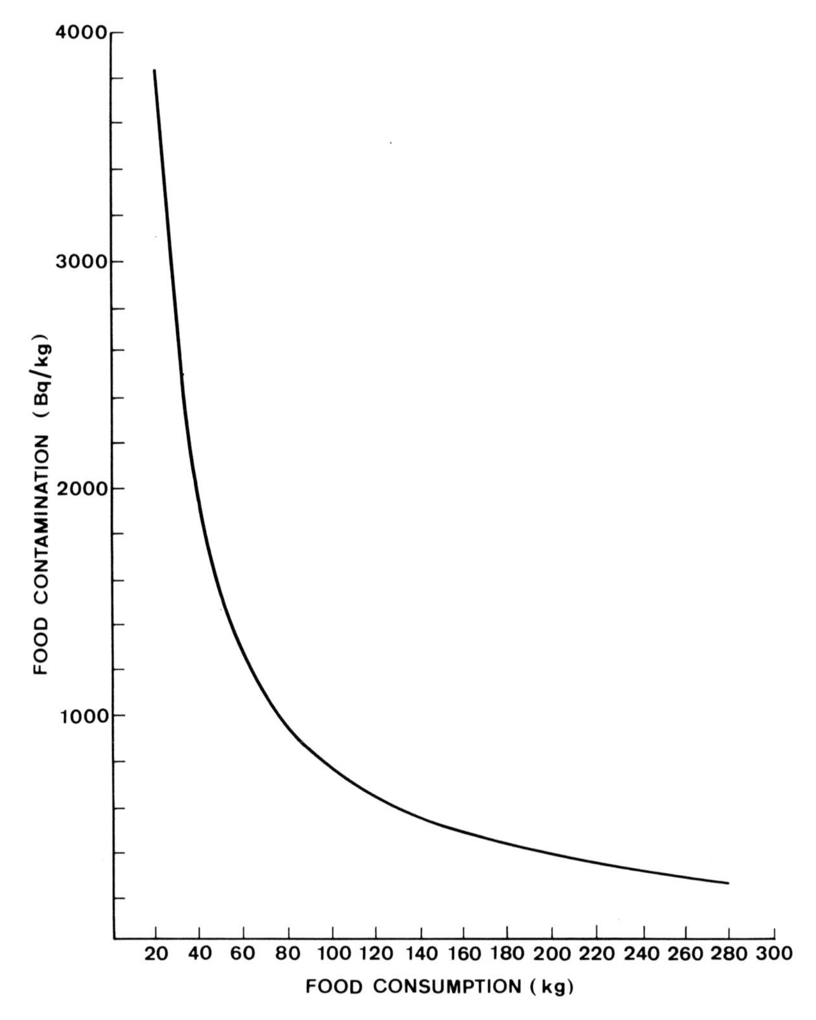

Estimate of food contamination, in Bq/kg, versus food consumption, in kg, for caesium-137, to give a committed dose of 1 millisievert assuming a dose conversion factor of 1.3×10^{-8} sievert/becquerel. Thus, for example, if the food contamination is 240 Bq/kg, then 280 kg of contaminated food would have to be eaten to receive a dose of 1 mSv (Waight, 1987).

reindeer belt escaped significant radiation levels. This disaster for the Lapplanders was the subject of a feature article "The Last Round-up? A Cloud Hangs over the Lives of the Lapps—Europe's Last Nomads", in the colour magazine of the *Sunday Times*, 30 November 1986. It stated that when 1000 reindeer were slaughtered for meat in September, in 97%, radioactive contamination of up to 10,000 Bq/kg was found, compared to the Swedish limit for meat consumption of 300 Bq/kg (the European Economic Community limit is 600 Bq/kg.) The initial government instruction to slaughter and bury the carcasses in pits has now been modified, so that some meat will be fed to foxes and minks in Sweden's fur farms, since these animals have no place in the human food chain.

As for the Soviet Union, their delegation papers at the 25–29 August 1986 meeting contained relatively little information concerning food contamination levels, distribution of stable iodine and food restrictions. Potassium iodate tablets were distributed to all the power plant workers at 0300 hours on 26 April, and in the town of Pripyat the tablets were distributed later that day at 2000 hours. This was undertaken by medical staff and local volunteers, who organised a door-to-door visiting schedule. Action levels for the Soviet authorities were 30 rem to the thyroid of children and 5 rem to the whole body of adults. Instructions to the population were linked to these levels. Thus milk restrictions were imposed on 1 May and other (unspecified) food restrictions, the first general restrictions (unspecified), were on 8 May, and these were revised and extended on 30 May.

The World Health Organisation organised consultation meetings on 6 May in Copenhagen and on 25–27 June in Bilthoven, The Netherlands. Following the Copenhagen meeting, there was also a limited circulation of a document entitled "Updated Summary of Data Situation with Regard to Activity Measurements, 12 June 1986", and contained data for thirty-five different countries. Information on iodine-131 levels in milk, and recommendations concerning the drinking of milk, have been extracted from this WHO summary document and are reproduced below. They show a wide variation, not only in activity levels, but in what actions are recommended. This was a feature of post-Chernobyl recommendations in Europe.

Albania

Limit for fresh milk is 2000 Bq/l. All measured levels less than 800 Bq/l (5 May).

Austria

Population will only be provided with fresh milk which has been checked before delivery (5 May). Milk above 10 nCi/l (=370 Bq/l) is not put on to the market (6 May).

Belgium

Previously recommended measures of keeping the milking cows indoors has been withdrawn (15 May). Concentration in farm milk reached a maximum of 85–170 Bq/l on 8 May but by 10 May decreased to 40–80 Bq/l. On 11–15 May it was 28–57 Bq/l.

Bulgaria

Consumption of sheep's milk banned as iodine-131 concentration "above normal" [9 May]. Temporary accepted contamination levels for milk are 2000 Bq/l for adults and 500 Bq/l for children. On 19 May the concentration in farm milk was about 100 Bq/l. Consumption of sheep's milk was still banned on 13 May.

Canada

Precautionary limit of 10 Bq/l instituted for precipitation used as the sole source of drinking water and the same limit applied to milk (14 May).

Czechoslovakia

Concentrations first reported to be of the order of 500 Bq/l for dairy milk and up to 1000 Bq/l for farm milk (4–5 May). The highest recorded concentration was 1570 Bq/l on 11 May. On 13 May provisions were made to stop direct consumption of sheep's milk and fresh products made from it. An accepted limit of 1000 Bq/l was established for the sale of cow's milk.

Denmark

Milk showed only slight contamination (28 April). Levels in milk far below the limits acceptable for human consumption (30 April).

Finland

In southern Finland levels were 10–40 Bq/l on 30 April. Cows can now be let out to pasture also south of the line from Kokkola to Kajaani and it is predicted that the action limits of 200 Bq/l iodine-131 and 1000 Bq/l caesium-137 will not be exceeded in the near future (26 May). Since the start of the grazing season on 26 May, the concentration of iodine-131 and caesium-137 has remained below 10 Bq/l (3 June).

France

The French Government adopted a control level of 2000 Bq/kg iodine-131 for milk, dairy products, fruit and vegetables. On 20 May, concentrations measured in dairy products were in the range 80–110 Bq/kg, and on 4 May a measurement gave 5.4 Bq/kg.

German Democratic Republic

Regular determinations of radioactivity have shown no need for special public health actions (2 May).

Federal Republic of Germany

Measured concentrations of 150–600 Bq/l in Munich (1 May). Milk contamination "increasing somewhat" (5 May). Maximum value of 500 Bq/l was reported (14 May). Iodine-131 in milk is generally below 250 Bq/l with caesium-137 concentrations up to 300 Bq/l (16 May). Import limits for the period 1 June to 30 September 1986 for caesium-137 (there has been a control of caesium-137 in foodstuffs since 1957) are 370 Bq/l for milk and dairy products and for baby food, and 600 Bq/l for other foodstuffs.

Greece

Recommendations since 5 May to avoid the use of fresh milk, mainly the milk which comes from sheep and goats (13 May). On 10 May the measured content in cow's milk was 100–400 Bq/l, in sheep/goat's milk was 2000–8000 Bq/l and in dairy milk 150 Bq/l.

Hungary

Public advised to consume milk marketed by the State Milk Industry in prepackaged form. Farms have been advised to discontinue pasturing and to feed animals on stored fodder (6 May). Dairy milk prepared as a mixture of fresh milk collected from both contaminated and uncontaminated areas, which is controlled for radioactivity before distribution. Limit for direct consumption of fresh milk and milk products is 500 Bq/kg. Actual levels in prepacked milk are below 150 Bq/l for iodine-131 and below 20 Bq/l for caesium-137 (15 May). Iodine-131 levels in farm milk, Bq/l, for cows on pasture, were 100–700 on 1–2 May, up to 1250 on 3 May and up to 2600 on 4 May (these compare with 100–200 and 200–800 on these two days for cows not on pasture); maximum levels around 1000 to 1500 up to 9 May, after which date they begin to fall, measuring 700 Bq/l on 13 May.

Iceland

No increased levels of radioactivity found (2–5 May).

Ireland

It was not necessary to institute any public health measures. During May, the mean measured level for iodine-131 was 21 Bq/kg, and the maximum, which was measured in lettuce, was 140 Bq/kg.

Israel

No precautionary measures were necessary. Limit set for milk was 2000 Bq/l, but on 3 May the maximum measured was 0.7 Bq/l. Goat's milk was found to be about 22 Bq/l.

Italy

Instructions given that babies, children under 10 and pregnant women should not drink fresh milk (2 May). Concentrations in dairy milk for 2–8 May were, in Bq/l, 55–300 for northern Italy, 35–185 for central Italy and 7–550 for southern Italy. Peak levels were later reported as 3000–6000 Bq/l in northern and central parts of the country. Values fell by factors of between 2 and 5 respectively by 15 May.

Japan

Contamination in the range 0.4–3 Bq/l (4–6 May).

Luxembourg

No public health measures necessary (2 May).

Malta

13 Bq/l on average for dairy milk with 140 Bq/l maximum (13 May).

Monaco

No public health measures necessary.

Netherlands

Fresh sheep milk excluded from consumption and the production of sheep cheese to be consumed within 5 weeks of preparation is forbidden. No regulations with regards to goats, since they are kept stabled and are not fed fresh grass (4 May). Forbidden to allow cattle out of doors, or to graze, and it is expected that this regulation will be in force for 1 week (3 May). Maximum concentration in farm milk was 173 Bq/l (4 May).

Norway

Action levels are for iodine-131, 1000 Bq/kg, and for caesium-137, 300 Bq/kg. Peak level in milk was 30 Bq/l.

Poland

For 29–30 April, levels of between 30 and 2000 Bq/l were reported for milk in various areas of Poland, but by 11 May the range was 80–474 Bq/l. The first increase in air radioactivity had been detected at 2100 hours on 27 April by Polish radiation monitoring services. On 5 May, limitations regarding consumption of fresh milk were being continued and in certain regions cows were not allowed to graze outside. Polish limit for milk contamination with iodine-131 is for children 1000 Bq/l.

Portugal

Peak milk contamination in milk was 0.1 Bq/l of iodine-131 (7 May).

Romania

Contamination reported at 450 Bq/l of iodine-131 and 10 Bq/l of caesium-137 in milk. Raw farm milk with concentrations in excess of 1000 Bq/l were used for "adequate industrial processing". Established limits for milk were 185 Bq/l for children and 1000 Bq/l for adults.

San Marino

Consumption of fresh milk was allowed because cows are fed with stored fodder.

Spain

Levels in dairy milk were 0.3–1.8 Bq/l and in local farm milk were 2–65 Bq/l (5–7 May).

Sweden

Cows should be taken indoors until further notice (4 May). The Soviet Union's import prohibition of 30 April, which was for meat, fish, potatoes and vegetables, was extended on 5 May to include milk products, such as cheese, and fresh fruit. Measurements on mother's milk from the Stockholm region (27 April–4 May) showed 8–25 Bq/l. Deposition of iodine-131 varied between 6000 and 170,000 Bq/m^2, with the highest values in northern Sweden (caesium-137 deposition varied between 300 and 33,000 Bq/m^2). It was established that a deposition level of 10,000 Bq/m^2 of iodine-131 corresponded to an expected milk concentration of 2000 Bq/l. Iodine-131 concentration in milk varied from 2–70 Bq/l, except on the island of Gotland where 700 Bq/l was measured. In raw milk on this island, 2900 Bq/l was measured.

Switzerland

The following recommendations were issued: for children under 2 years, pregnant women and nursing mothers, the use of dried or condensed milk, packed before 3 May, is recommended; for the remaining population, free consumption of milk and milk products; refrain from drinking milk from sheep which have been grazing in the open. Initial concentrations 250 Bq/l for cow's milk, but this rose to 1370 Bq/l on 13 May. On 3 May, 5800 Bq/l for sheep's milk and 550 Bq/l for goat's milk.

Turkey

No danger is reported. 360 Bq/l in fresh milk and 48 Bq/l in dairy milk (6 May).

USSR

No information on milk or foodstuffs given in this WHO survey (see Chapter 6 for information on food restrictions from sources such as Novosti Press Agency and the Soviet delegation documents of 25–29 August 1986).

United Kingdom

Levels in milk are below the recommended levels at which restrictions on milk supplies would be considered in the United Kingdom (9 May). Range of 3 to 240 Bq/l for dairy milk and for farm milk a maximum of 370 Bq/l (2–5 May). Maximum measured milk concentration was 1136 Bq/l.

USA

Milk monitoring did not detect any activity in any sample (4–7 May). Import regulations for iodine-131 in infant foods is 56 Bq/kg.

Yugoslavia

For the period 2–31 May, most measured concentrations did not rise above 400 Bq/l.

The iodine-131 problem is largely related to milk, whereas the caesium-137 problem is largely related to various food chains. Because of its long half-life of 30 years, this radioactive isotope gave the highest contribution to population exposure from atmospheric nuclear weapons testing in the 1956–62 period. It has therefore been extensively studied by the United Nations Scientific Committee on the Effects of Atomic Radiation, UNSCEAR, which has published a number of reports and is due to report on the effects of the fall-out from Chernobyl in 1988.

UNSCEAR found from 1956–62 sources that caesium-137 contaminated most of the common foodstuffs such as milk, meat and cereals. However, one of the immediate worries of the health authorities and the population after Chernobyl was contamination of vegetables, particularly the leafy variety such as spinach, and fresh fruit. For instance, at the 6 May Copenhagen meeting, advice not to eat surface vegetables had already been given in Austria, San Marino and Sweden; and advice to wash fresh vegetables before eating has been given in Austria, Belgium, Federal Republic of Germany, Hungary, Japan, The Netherlands and Switzerland. By 12 June this had been amplified in the WHO summary with, for example, the following.

Austria

All green vegetables will be confiscated if grown in the open (6 May).

Greece

Vegetables showing a radiation level higher than average standards continue to be not allowed for human consumption, greater than 250 Bq/kg for iodine-131 and greater than 300 Bq/kg for combined caesium-137 and caesium-134 (23 May).

Italy

By 5 May the government placed a ban on the sale of leafy vegetables, but it was dropped after a few days in central and southern Italy, and after a few more days in the north.

Switzerland

The washing of vegetables such as spinach and lettuce had been recommended and it was reported that the washing procedure reduces the iodine-131 activity by approximately 30% and the caesium-137 by approximately 66%; and that if spinach was boiled, the reduction was even greater.

It is also interesting to note two papers in *Nature* by Hohenemser *et al.* (26 June) from the University of Konstanz in the Federal Republic of Germany and by Devell *et al.* (15 May) from Studsvik Energiteknik AB, Nykoping, Sweden. The German group found that fall-out aerosols brought down by rainfall showed a high resistance to weathering and drying and in experiments found that grass dried at 130°C for 24 hours, followed by a vigorous shaking, lost less than 10% of the adhering radio-activity. Cutting, drying and storing grass for winter cattle feed may therefore lead to radioactive accumulation in silos and barns. This problem of hay accumulation and storage may be all the more serious in view of the discovery by Devell *et al.* who found that fall-out aerosols contain 1–2 μm "hot" beta-emitting particles of strengths 1000–10,000 Bq activity. If these are widespread constituents of Chernobyl fall-out, they can be expected to be trapped permanently in the lungs of individuals exposed to the contaminated hay.

Materials other than food were tested. In the Federal Republic of Germany a test on sand was carried out in order to find a safe limit if a child playing in a sandpit ate 1 kg of sand—something one might have thought was a physical impossibility! Also in one national radiation laboratory in a European country, masses of redcurrants were brought to be tested, and virtually all were found to have negligible amounts of radioactive contamination. However, the fruit was not wasted and I have it on good authority that the home deep freezers of the staff of this laboratory are now full of redcurrant pies, redcurrant preserves and redcurrant trifles!

In 1987, the earliest European forum for discussion of the effects of Chernobyl, including consequences for the food chain, was a meeting on 8–9 January 1987 of the European Members of Parliament. This was reported in *Le Monde* on 13 January with the title "Une reunion du Conseil de l'Europe sur la catastrophe de Tchernobyl", and an English translation of some of the features covered is given below.

> *"The European parliamentarians met in Paris, 8–9 January, on the initiative of the Council of Europe (this is an organisation of twenty-one countries with a much larger membership than the EEC) to have explained the implications of Chernobyl. They must have returned to their countries more troubled than before. Mr Boris Semionov of the Soviet government had just foretold that the RBMK reactors not only will not be dismantled, but their numbers will be increased further by three units which are at present under construction. The only consolation for the Austrian and Scandinavian MPs who fear this dangerous proximity was that in the future the plan in the Soviet Union is to install PWR and not RBMK plants. Coming to the aid of his Russian colleague, the American Mr Morris Rosen of the IAEA underlined that in normal working conditions 'nuclear power stations are a lot more safer and cleaner than the classic thermal power stations'. According to Rosen, there would be every year in the United States some 10,000 deaths by natural radiation and more than 2000 fatal cancers due to medical radiation. This is relatively low when compared with the 14,000 deaths from the classic electricity production cycle, with the 17,000 deaths from firearms, with the 50,000 deaths from road accidents, with the 100,000 victims of*

*alcohol, and with the 150,000 deaths 'probably' due to tobacco. 'The accident at Chernobyl is not acceptable, but it is tolerable', Rosen repeated. The European MPs were hardly appreciative of these comparisons. As for those who wanted precise and practical answers, their hunger for knowledge remained unabated. What distance from the accident did farm animals die? asked a Scottish MP concerned with lamb. 'The Russians have told us nothing on this subject', replied the expert. 'And the pollution of the water?' asked an English MP. 'Impossible to give a general response', declared the Spanish expert, 'certain radioactive elements have been absorbed by the soil, others by streams, everything depends on the ecosystem and the particular radionuclides'. A Luxembourg MP was astonished by the disparities in planned zones of evacuation. In France it is 10 kilometres, in the United States it is 10 miles (16 kilometres), in Germany and Switzerland it is 20 kilometres, and at Chernobyl they evacuated to a distance of 30 kilometres. An English MP insisted 'but what is the level of becquerels acceptable per kilogram of meat?' 'There is no common level in Europe', declared the Director of the Swedish Institute of Radiation Protection, Mr Bengtsson.**

"Chernobyl is the name in Russian of a variety of absinth."
(*Le Monde*, 7 January 1987)
(Chernobyl has also been said to mean "black wormwood".)

---

* See note 7, p 234–235.

8

The Entombment of the Reactor

> "The Soviet term for the entombment is 'Sarcophagus'. The earliest stone sarcophagus belongs to the Egyptian third dynasty which ended in 2620 BC. A description of this sarcophagus/tomb/coffin is 'A rectangular box of white limestone with plane sides and a slightly vaulted lid.' "
>
> (*Death in Ancient Egypt*, A. J. Spencer, 1982)
>
> This description almost fits the shape of the very much larger Chernobyl sarcophagus more than 4500 years later.

IT HAS already been mentioned in Chapter 3 (2100 hours on 26 April 1986) how immediately after the accident decisions had to be made as to the best way of stopping the graphite fire in the reactor core and that in the end the choice was "to contain the accident at source by covering the reactor shaft with heat-absorbent and filtering materials". The Soviet delegation, 25–29 August 1986, baldly described this in the following terms: "A group of specialists began to cover the damaged reactor by dropping compounds of boron, dolomite, sand, clay and lead from military helicopters. About 5000 tonnes in all were dropped between 27 April and 10 May, mostly between 28 April and 2 May. As a result, the reactor was covered with a friable layer of material which strongly absorbed aerosol particles."

In practice, the operation, which was in the charge of Major General Antoshkin of the Air Force, took hundreds of helicopter flights before, in the language of Novosti, the reactor was "reliably choked up with a huge flaky pie of sand and other materials". The first "crater bombing missions" were the most difficult and on the first day ninety-three missions were flown and on the second day 186 missions. Overflying speed was 140 kilometres per hour and the accuracy of the sandbag drop was aided by instructions from a monitoring plane equipped with instruments. Initially, only a single sandbag of material was dropped each flight, but at the suggestion of the helicopter pilots, special packages were invented. These were six to eight sandbags tied together, using makeshift nets so that the pilots were able to throw such a package through the helicopter hatch, having themselves designed a self-opening lock. A photograph of one of these packages, taken through the hatch of a helicopter, was published in *Pravda* on 20 May and is reproduced in Fig. 24. Only when the reactor crater was sealed was a TASS correspondent allowed to overfly and take pictures. Earlier photographs were all taken by the military or by scientists.

The mix of materials dropped by the helicopter pilots was chosen for specific purposes. The boron was to absorb neutrons and to stop any possibility of the reactor becoming critical again (that is, starting up a fission chain reaction); the lead absorbed heat, melted into gaps and acted as shielding; whilst the sand acted as an efficient filter. The dolomite gave off carbon dioxide as it heated up and this reduced the flow of oxygen to fuel the graphite fire. Some of this material was air-freighted into the Soviet Union from various countries, and at least several tons of silicates left from London's Heathrow, although this was not well advertised at the time. Some countries contributed equipment to help contain the effects of the accident, and an example was a number of remote-controlled robots, developed in the Federal Republic of Germany (See Figs. 86, 87, 88). One design was an agile instrument which could measure radiation levels throughout the plant. It was a miniature vehicle equipped with stereo television cameras, microphones, radiation monitors, temperature sensors, sampling devices, still cameras and a variety of specialised interchangeable manipulator arms and tools. A larger type of robot was also provided by the Federal Republic to assist with demolition and removal of reactor debris and equipment. It could also undertake excavation.

About 1 May the temperature of the nuclear fuel started to rise due to heating caused by the radioactive decay of nuclear fission products remaining in the damaged reactor and by graphite combustion. To solve this fuel temperature problem, nitrogen was pumped under pressure from the power plant's compressor station into the space beneath the reactor vault and by 6 May the temperature in the vault had begun to decrease. The maximum temperature[8] of the fuel during this period was 2000°C.

An underground bunker was dug under one of the buildings of the power plant, some 600 metres from unit No. 4, and acted as the control outpost for the co-ordination of site operations at the plant. Most of the work was carried out by army personnel, but a special group of coal miners from Tula and the Donets basin were also drafted for the emergency. Their particular role was to build a "cooling slab" under the reactor of unit No. 4. (See Fig. 29.) This was a form of insurance (which in the event was not needed) against the lower structures of the reactor building being destroyed. It consisted of a flat reinforced concrete slab incorporating a flat heat exchanger. Limited working times of 3-hour shifts were set for the miners because of the conditions, and the entire tunnelling operation which involved more than 400 people was completed on 24 June. The first few metres were the most difficult, as the tunnellers had to drive to a depth of 6 metres in solid sandstone, constantly monitoring radiation levels all the while. This was started after first digging a large pit near unit No. 3. The tunnel, which is encased in reinforced concrete, is 168 metres long and 1.8 metres in diameter, with the final 5–6 metres worked manually. The rate of tunnelling was 60 centimetres of sandstone rock per hour, with the tunnel being completed in 15 days. Service lines and rails for buggies were laid inside and thirteen small galleries were dug off the tunnel. In these galleries, riggers assembled what were called dampers, which are devices for cooling off the reactor's foundations. A monolithic reinforced concrete slab was finally installed underneath the damaged reactor.

The plans for the long-term entombment of the reactor were described by the Soviet delegation, 25–29 August 1986, as an intention to build the following engineering structures which they termed a "sarcophagus":

Outer protective walls along the perimeter.

Inner concrete partition walls in the turbine hall between units Nos. 3 and 4, and in other places, including alongside the debris of the tank room of the emergency cooling system.

A metal partition wall in the turbine hall between units Nos. 2 and 3.

A protective roof over the turbine hall.

It was also proposed to seal off the central hall and other reactor rooms, and to pour concrete over the debris by the tank room and over the rooms of the northern main circulation pumps. This would isolate the debris and provide radiation protection from the reactor sector. The thickness of the concrete walls were to be 1 metre or more, depending on radiation dose rates.

Ventilation systems[9] with heavy duty filters and radiation monitoring devices were installed within the entombed reactor and by 14 July a device called "The Needle" had been inserted. This is an 18 metre steel pipe, 10 centimetres in diameter, painted black and white and packed inside with instrumentation to measure temperature and radiation levels. Three helicopters were used for its insertion into what TASS and Novosti often call the "sarcophagus" of unit No. 4; it took three attempts. It is linked to a 300 metre fall-line containing cables to transmit data from instruments not only in the pipe itself, but also from those installed under an "umbrella" put at the joint between pipe and fall-line. The Needle was driven inside the sarcophagus so that two-thirds were inside and one-third outside, with the umbrella left over the reactor. The fall-line was dropped in such a way that it lay along a wall of the reactor where a communication line was linked to it for picking up continuous readings from the instruments.

Many of the workers involved in these operations lived on floating hostels and hotels in the River Pripyat. Convoys of ships had sailed from various Soviet ports in order to help provide temporary facilities. One such convoy sailed from the city of Azov, crossed the Sea of Azov, the Black Sea, and then entered the River Dnieper on its way to the power plant. Another convoy consisted of fourteen ships from the Volga and Kama, which travelled via the Volga–Don shipping canal and the southern seas. These included eleven passenger ships to act as hostels and hotels. Other vessels included drinking water carriers, water treatment plants, and a floating shop.

The sequence of decontamination, wall construction and entombment activities was (USSR Delegation, 25–29 August 1986):

1. The surface layer of soil in the area adjacent to unit No. 4 was removed to local waste disposal sites by special remote control technology. (When crane operators had to control their machines manually, lead-plated cabins were provided for the cranes.)
2. The area was covered with concrete and the surface levelled to facilitate the movement of self-propelled cranes and other machinery.
3. The roofs and walls of buildings were decontaminated.

4. After the site was cleaned and covered with concrete, the metal frames for the protective walls were assembled and then covered with concrete.
5. As the walls were being built, work proceeded with the construction of the main civil engineering structures which were to ensure entombment of unit No. 4.

Final entombment occurred in mid-November, somewhat delayed, due to there being a shortage of concrete during operations.[10] This was perhaps not surprising since, according to TASS, by 15 September already more than 160,000 cubic metres of concrete had been poured into shielding structures, which by then had reached a height of 41 metres. This also included the wall, which was really a trench 60 centimetres wide and 30–35 metres deep, that runs along the perimeter of the Chernobyl power station, and is filled with concrete up to the impermeable clay layer.

The power plant workers' satellite town of Pripyat will be replaced by the new town of Zelyony Mys* which is to be built within 2 years, on the bank of the River Dnieper near the Kiev reservoir, and a population of at least 10,000 is planned with a cinema, restaurants, medical institutions and a stadium being provided. A widened highway to the nuclear power plant has already been completed so that access is quicker than before the accident. At present, the workers are in hostels and hotels, such as those mentioned earlier, and the families of the power workers are living in Kiev and Chernigov.

---

* When translated this means "green cape".

FIG. 116. A house in the Zhlobin settlement.

Courtesy of Novosti

FIG. 117. A settlement in the Zhlobin area of the Gomel region of Byelorussia.

Courtesy of Novosti

FIG. 118. Traditional bread and salt is presented to the family of an evacuated tractor driver outside their new home, September 1986.

Courtesy of TASS

FIG. 119. It took construction workers from Ternopol 50 days to build Ternopolskoye, a village of 150 homesteads in the Makarov district, for the members of the Chervone Polissya state farm evacuees. Exterior decoration of the houses, weather vanes, dovecotes, letter boxes, shelves in the cellars and some other items were not provided in the official construction project. The building workers added them on their own initiative, September 1986.

Courtesy of TASS

FIG. 120. A red flag is hoisted on a crane to celebrate the achievement of an interim goal in the clean-up operations – filling the gap in the wall and sealing the roof of the damaged unit No. 4, October 1986.

Courtesy of TASS

FIG. 121. The three men who were recognised as the best three workers at Chernobyl. *From left to right*: Muscovite V. Butkov, Deputy Director of the Chernobyl Construction Board; construction site manager, S. Zykov who came from Barnaul; and from Uzbekistan, M. Zangirov, a carpenter and concrete worker.
Courtesy of TASS

FIG. 122. The team leader of a radiation control group measuring the radiation levels after decontamination of the area, October 1986.
Courtesy of TASS

FIG. 123. A floodlit view of the ribbed wall of the sarcophagus, December 1986.
Courtesy of TASS

FIG. 124. A close-up view of a concrete wall reinforced with steel structures and outward ribs, December 1986.

Courtesy of TASS

FIG. 125. The concrete entombment in December 1986. The construction workers were drawn from many parts of the Soviet Union and included volunteers from Moscow, Tomsk, Krasnoyarsk, Yuzhnouralsk and Shevchenko.

Courtesy of TASS

FIGS. 126 & 127. The press conference held on 15 January 1987 at the Literary Cafe in Dzerzhinsky Street, Moscow, attended by Lieutenant Colonel Telyatnikov, now honoured as a Hero of the Soviet Union. On his left is Professor Angelina Guskova, Head of the Clinic of the Institute of Biophysics and Section Head at Clinical Hospital No. 6. It was Professor Guskova who was the medical doctor in charge of the treatment of the 56 most desperately ill victims of the accident who were evacuated to Moscow Hospital No. 6, and who worked on the bone marrow transplants with Dr Robert Gale. In an 18 September 1986 interview with *Izvestia*, Professor Guskova reported that after the accident 300 people were hospitalised, 203 were diagnosed with acute radiation sickness and 31 (all of them either Chernobyl nuclear power plant workers or firemen) had died. However, by the time of the interview, Oleg Shchepin, Soviet First Deputy Health Minister, was able to report that only 3 patients remained in hospital in Moscow and only 11 in Kiev. By January 1987 it was believed that fewer than a total of 5 patients were still in hospital.

Courtesy of TASS

КОЛОКОЛ ЧЕРНОБЫЛЯ

BELL OF CHERNOBYL

FIG. 128. A Soviet Central Documentary Film Studio film with the title "The Chernobyl Warning Bell" was premiered on 3 March 1987 at the Oktyabr Cinema in Moscow. It included a series of interviews with nuclear engineers, firemen, helicopter pilots and builders who took part in the clean-up operation, as well as interviews with local farmers and Pripyat residents, and with Armand Hammer. One sequence is of the plant's chief engineer Nikolai Shteinberg conducting a tour of the roof covered with radioactive debris. The accompanying soundtrack includes his warning: "There round the corner, 10 rontgens, the cover over there (3 metres from the camera) is nearly 200. That pipe there (10 metres distant) is roughly 1,000 rontgens, OK lets get away quick." It is not clear what is the exposure-rate but the total radiation exposures quoted would obviously be received in a short period of time, certainly not measured in hours. This is emphasised by the soundtrack referring to a short briefing for servicemen before they commence a cleanup procedure: "As you leave this step, start counting to 90, then drop your barrows and spades and sprint back." This cinema film was preceded on 18 February 1987 by a Soviet Central Television showing, 80 minutes in length, of a documenty film "The Warning", which used film sequences shot by amateur cameramen (presumably military personnel in helicopters), who had been on site immediately after the reactor was damaged. This includes a sequence showing part of the reactor's graphite core glowing red hot, and shots of the miners tunnelling under the reactor and of helicopter pilots and their lead/silicates/dolomite loads read for the unit No. 4 "bombing missions". The chief engineer, Nikolai Shteinberg also appears in "The Warning", and as Novosti comments, states that: "We are all to blame, including those who were away from the station at the tragic moment. The Chernobyl accident is the result of negligence, of the violation of regulations by several workshifts, not by one person."

There is also an admission (Novosti) that: "There were also panic-mongers, quitters and even marauders." This latter must refer to the Soviet TV interpreter's comment in "The Warning" that there were some looters in Pripyat during the evacuation. It has also been announced, recently, that some of those responsible will be taken to court and tried, but at the time of writing it is not known how many defendants there will be, nor their official positions in the power plant heirarchy. It should be remembered though, that by far the vast majority of those involved behaved in a responsible and courageous manner at the time of the accident and that panic-mongers and looters were very few and the admission that they existed should not be taken out of context.

The film "The Warning" has been purchased by the British Broadcasting Corporation and an abridged 40 minute version was shown on the BBC Panorama programme of 6 April 1987. I am most grateful to Soviet TV and to BBC TV for giving me a copy of the original film with Soviet soundtrack and for providing me with freeze-frame transparencies to be included in this book.

FIG. 126. *Courtesy of TASS*

FIG. 127. *Courtesy of TASS*

FIG. 129. Presentation of medals in the Kremlin on 14 January 1987 to those who participated in the events following the Chernobyl accident. TASS were not supplied with the names of the medal recipients, but 7th from the left is Mr Andrei Gromyko, President of the USSR Supreme Soviet Presidium, and second from right is Lieutenant Colonel L. Telyatnikov. It is thought that the man in military uniform on the right of President Gromyko is General I. F. Kimstach who was at the IAEA August 1986 meeting and who is the head of the Main Fire Protection Directorate of the Ministry of Internal Affairs. Others who have received medals for their part in the Chernobyl aftermath were mentioned in a Novosti report of 25 December 1986. These include the title Hero of the Soviet Union awarded to Air Force Major General N. T. Antoshkin (whose part in the helicopter operations to entomb the reactor have been mentioned in Chapter 8) and to Colonel General V. K. Pikalov of the Chemical Defence Force (whose experiences have been described in the captions to Figs. 90 & 96). The Order of Lenin and the Hammer and Sickle Medal was presented to: V. I. Zavedy, leader of the team whose concrete laying beat all records; G. D. Lykov, chief of the construction task force that buried the damaged unit; Y. N. Samoilenko, who was chief of the team which decontaminated the roof of the damaged unit and who is now the deputy shop superintendent of the Rostov nuclear power plant which is under construction; and A. N. Usanov, Deputy Minister of Medium Machine Building of the USSR, who was on site from the first days after the disaster and was responsible for arranging a three-shift system which made it possible for the unit to be buried within the projected time schedule.

Courtesy of TASS

FIG. 130. Major (now Lieutenant Colonel) Leonid Telyatnikov being awarded the Order of Lenin and the Gold Star medal on 14 January 1987.

Courtesy of TASS

ЗАДАНИЕ ПРАВИТЕЛЬСТВА ВЫПОЛНИМ

FIG. 131. Announcement of completion of the entombment as it appeared in *Pravda*, 15 November 1986.

САРКОФАГ

SARCOPHAGUS

FIG. 132. In the Soviet Union in the 1920s plays about recent historical revolutionary events were dramatised and taken round the country by groups of actors. This practice fell into disuse for many years but has been revived after the Chernobyl accident with a play called *Sarcophagus: A Tragedy*, written by the science correspondent of *Pravda*, Vladimir Gubaryev.* Performed in the USSR soon after its publication in June 1986, its premiere outside the Soviet Union was on 9 April 1987 by the Royal Shakespeare Company at the Barbican Theatre in the City of London. The word Chernobyl is only mentioned in the dedication, but there are obvious similarities to the actual events in the Ukraine in the list of the cast, variously described as professor of surgery, physician and research scientist, medical director of the Institute of Radiation Safety, official of the State Prosecutor's Department and American professor of surgery. Among the irradiated are a cyclist, fireman, driver, Director of Nuclear Power Station, Geiger counter operator, control room operative, general and physicist. The cyclist might seem an odd inclusion, but Michael McCally who visited Moscow Hospital No. 6 on 6 June 1986 and interviewed patients describes those with radiation sickness as "Of the 299 patients who were admitted, most were men aged 25 to 35 and two were women, 10% were over 40 years old. With two exceptions, all were technicians at the power station or rescue workers. Two Chernobyl residents who had been walking or bicycling nearby required hospital treatment." The final speech of *Sarcophagus* is a male voice over the radio making an announcement from the cast, director and author: "This play is dedicated to Pravik†, Lelechinenko, Kibenok† and Ignatenko†, Tishchura† and Vashuk†, Titenok† and Telyatnikov, Busygin and Gritsenko, to the firemen and power station workers, the physicists and calibrators, the officers, helicopter pilots and miners, adults and children – to all those who, at the cost of their lives and health, extinguished the nuclear flames of Chernobyl."

The advertising describes the intentions of the playright in the following terms: "Gubaryev is highly critical of the Soviet system, but he also examines the relationship of the West to the USSR and asks searching questions about our attitudes to cover-ups and secrecy. He uncovers alarming facts about the inadequate safety measures in the nuclear plant, implying that these faults were known about in advance by the management of the power station. But by setting *Sarcophagus* in a radiation clinic, where the victims were sent after the disaster, Gubaryev also shows the effect that Chernobyl had on the local population."

---

* One of the surprises of my 2 December 1987 visit to Chernobyl was the vehemence with which Mr Alexander Kovalenko of "KOMBINAT" assailed Gubaryev for "distorting events and writing only 40% of truth". There is now a heated debate in existence in the USSR press about how far writers should go in using invented characters and events to dramatise the Chernobyl accident for public consumption. Inventions of a "cyclist–burglar" and the partly crazy "alcoholic scientist Fred the undead" were on one side claimed to be acceptable artistic licence, whereas on the other side they were seen as gross departures from the truth. Be that as it may, though, "Fred" was certainly good theatre in the Royal Shakespeare production – and who knows, his doppelganger may exist somewhere!

† These are the six firemen whose obituaries were given in the 19 May 1986 issue of *Izvestia* (see Fig. 58).

FIG. 133. A scene from the Royal Shakespeare Company production of *Sarcophagus*, April 1986. On the left is a patient called Mr Immortal who has been in the special radiation injuries unit for 18 months after a plutonium accident. This, according to the play, occurred when he had drunk too much vodka, fell asleep in the laboratory where a plutonium experiment was underway, and was irradiated for 3 hours whilst recovering from his hangover. He was also called "Fred the undead". On the right, with back to camera, is the professor in charge of the radiation victims. She is based on Professor Angelina Guskova.

Courtesy of Michael le Poer Trench

FIG. 134. Photograph of the author and Colonel Telyatnikov at the Soviet Embassy, London, 6 March 1987. At the invitation of the *Star* newspaper, Colonel Telyatnikov, together with the Head of the Soviet Fire Services, General Anatoly Kuzmich Mikeev, visited London from 4 to 7 March 1987. On the 5th he was given an award by the *Star* and visited No. 10 Downing Street to meet the Prime Minister. On the 6th, at a small Soviet Embassy press conference attended by American TV networks and some of the British media, the British Fire Services Association presented Colonel Telyatnikov with their highest award, the Order of Gallantry, which he accepted "as an appreciation of all the firemen of Chernobyl and of those no longer with us". I was fortunate to be invited to this ceremony to be introduced to this very brave fireman and given the opportunity to talk to him for a few minutes. At the press conference, held some 10 months after the accident, he looked extremely well with his dark brown hair completely regrown after having had it shaved off immediately on admission to hospital. When asked by the press if he had experienced any radiation sickness he answered "Yes" and described his case history since the accident as "2 months treatment, 1 month rehabilitation [see Fig. 106], 1–2 months health resort treatment at a sanatorium [see Fig. 105], 2–3 months of general medical checks, and that he was now cured of radiation sickness and had, with his colleagues, resumed his firefighting duties some $1\frac{1}{2}$ months ago." He was also asked what were his most difficult moments, how did he hear of the disaster, what were his first reactions and what did he do when he arrived at the power plant. The reply was as follows, and mentions two of the firemen who eventually died [see Fig. 58]. "The first report of the accident was at 0123 am on 26 April 1986 by an alarm of the control [panel] in the Chernobyl firefighting unit. The unit on duty was that of Lieutenant Pravik and was the first to go to the scene. Then some 5 minutes later Lieutenant Kibenok and others arrived. I was at home at the time of the alarm and rushed to the plant to take charge. On arrival, I noticed other firefighters already engaged in firefighting. What was most remarkable, what I will always remember, was the way in which all colleagues were acting bravely and with self-sacrifice. What was probably the most difficult part was the knowledge that all knew of their responsibility and of the radiation hazards, but they still continued heroically."

Courtesy of TASS

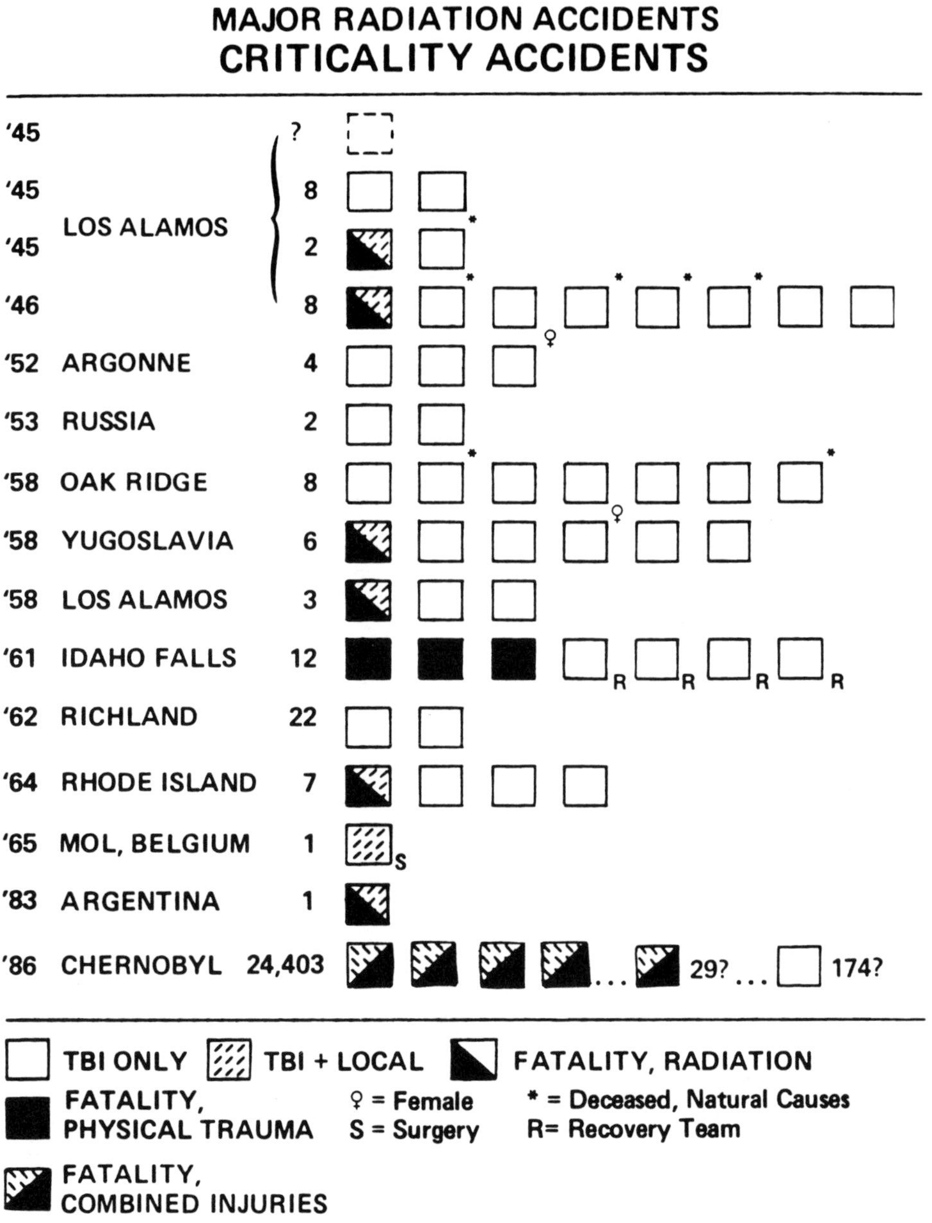

MAJOR RADIATION ACCIDENTS
CRITICALITY ACCIDENTS

Year	Location	Cases
'45	LOS ALAMOS	?
'45	LOS ALAMOS	8
'45	LOS ALAMOS	2
'46	LOS ALAMOS	8
'52	ARGONNE	4
'53	RUSSIA	2
'58	OAK RIDGE	8
'58	YUGOSLAVIA	6
'58	LOS ALAMOS	3
'61	IDAHO FALLS	12
'62	RICHLAND	22
'64	RHODE ISLAND	7
'65	MOL, BELGIUM	1
'83	ARGENTINA	1
'86	CHERNOBYL	24,403 (... 29? ... 174?)

TBI ONLY — TBI + LOCAL — FATALITY, RADIATION

FATALITY, PHYSICAL TRAUMA

♀ = Female

S = Surgery

* = Deceased, Natural Causes

R= Recovery Team

FATALITY, COMBINED INJURIES

FIG. 135. A diagram of medical cases recorded in the REAC/TS radiation accident registry who were victims of criticality radiation accidents in which fissionable nuclear materials were involved.

Courtesy of C.C. Lushbaugh

FIG. 135. On 2 April 1987 the British Institute of Radiology held a seminar entitled "Nuclear reactor accidents: preparedness and medical consequences" and I am grateful to Dr C. C. Lushbaugh of the U.S. Department of Energy, Radiation Emergency Assistance Center and Training Site (REAC/TS) in Oak Ridge for permission to publish these two figures (135 and 136) which he presented at the seminar and which will be published with the proceedings in the *British Journal of Radiology*. (Reference: Lushbaugh, Fry and Ricks, 1987.)

This diagram shows the 43-year history of acute radiation deaths in major radiation accidents, by year, location and number of persons involved. Each square represents a single individual and the shading notation represents the type of irradiation and the outcome.

The total number of radiation accidents in the REAC/TS Radiation Accident registry are as follows:

Number of accidents	Persons involved	Significant exposure	Fatalities (acute effects)	
284	1358	620	33	Before Chernobyl
1	135,000	24,200 (+203)	29	Chernobyl Accident

The term "Significant exposure" refers to those persons who received a total body irradiation (TBI) estimated as having been greater than 0.25 Gy or doses to the skin in excess of 6.0 Gy (620 before Chernobyl). For the Chernobyl accident, Lushbaugh *et al.* (1987) estimated the figure of 24,200 from Soviet published data. The figure of 203 is shown separately, since it refers to those who received doses greater than 1 Gy. They were all hospitalised and gave histories of acute gastrointestinal illness and vomiting within 6–24 hours of the explosion.

The term "Persons involved" refers to those who are thought to have had more than 0.01 Gy but less than 0.25 Gy total body irradiation (1358 before Chernobyl).

The term "Fatalities" refers to those caused by total body irradiation in combination with any or all of the following three injuries: thermal damage, beta radiation damage, toxic epidermal necrolysis.

Lushbaugh *et al.* (1987) have included in this diagram three radiation fatalities [Canada (1985), Georgia/Texas (1986)] caused by "computer error" in a radiation therapy computer-controlled linear accelerator used to treat cancer patients. "Computer error" is in fact another example of human error, this time with a new generation high technology medical linear accelerator, which unlike earlier generation machines are dependent on computer settings. The 29 fatalities from Chernobyl are those which are radiation deaths, but there were also two non-radiation caused deaths, making the total death toll equal to 31.

Courtesy of C.C. Lushbaugh

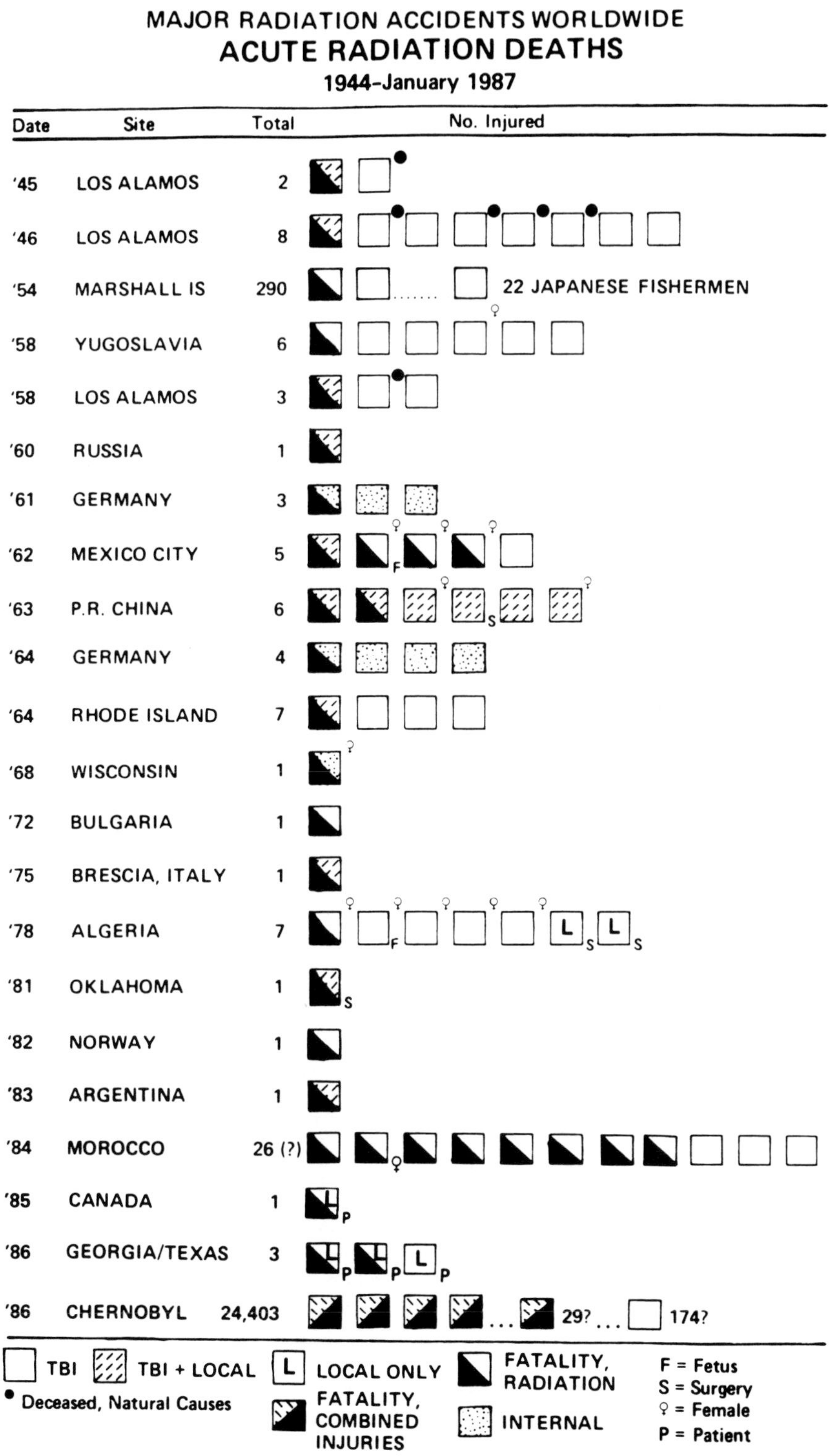

MAJOR RADIATION ACCIDENTS WORLDWIDE

ACUTE RADIATION DEATHS

1944-January 1987

Date	Site	Total	No. Injured
'45	LOS ALAMOS	2	
'46	LOS ALAMOS	8	
'54	MARSHALL IS	290	22 JAPANESE FISHERMEN
'58	YUGOSLAVIA	6	
'58	LOS ALAMOS	3	
'60	RUSSIA	1	
'61	GERMANY	3	
'62	MEXICO CITY	5	
'63	P.R. CHINA	6	
'64	GERMANY	4	
'64	RHODE ISLAND	7	
'68	WISCONSIN	1	
'72	BULGARIA	1	
'75	BRESCIA, ITALY	1	
'78	ALGERIA	7	
'81	OKLAHOMA	1	
'82	NORWAY	1	
'83	ARGENTINA	1	
'84	MOROCCO	26 (?)	
'85	CANADA	1	
'86	GEORGIA/TEXAS	3	
'86	CHERNOBYL	24,403	... 29? ... 174?

TBI; TBI + LOCAL; L LOCAL ONLY; FATALITY, RADIATION; FATALITY, COMBINED INJURIES; INTERNAL

• Deceased, Natural Causes

F = Fetus
S = Surgery
♀ = Female
P = Patient

FIG. 136. *Courtesy of C.C. Lushbaugh*

ONE YEAR AFTER BY TASS

A series of photographs of the Chernobyl nuclear power plant were taken by the TASS photographers V. Samokhotsky and V. Repik under the title *One Year After* (See Figs 137–142). Similarly, to mark the first anniversary of the accident, the Novosti Press Agency issued several articles of which the following four were translated (1 July 1987) into English: "The danger zone a year later", by Lev Voskresensky; "In the fields and farms of the Ukraine in 1987: the margin of safety", by Alexander Tkachenko, First Vice-Chairman of the Ukrainian State Agroindustrial Committee; "Babies born after the Chernobyl accident: how are they?", by Yelena Alikhanyan; and, "Problems before the Kiev Centre of Radiation Medicine", by Vladimir Kolinko. Much of the information in these articles already appears elsewhere in this book but additional data items not previously noted are as follows.

On radioactivity in river water

Results of Dnieper and Pripyat river water analysis for radioactivity concentration (presumably for radioactive iodine only):

Prior to April 1986	1×10^{-6} microcuries per litre
3 May 1986	3×10^{-2} microcuries per litre (maximum iodine concentration)
Mid-June 1986	1×10^{-4} microcuries per litre
May 1987	1×10^{-5} microcuries per litre

On the incidence of newborn twins

Professor Anatoli Zakrevsky, head of the Department of Obstetrics of the Ukrainian Institute of Mother and Child Care, Kiev, a 500-bed hospital, stated that: "The babies have no pathologies which might be due to the result of the accident. Laboratory tests show that babies born to women from the evacuated area (over 2000 were born by April 1987) are no different from those born before the accident. Peculiarly, 6% newborns are twins, a proportion above the average of 0.5%. I don't know whether it's a coincidence or pattern."

On radiation dose to Kiev population

Within 12 months (following the accident) Kiev residents received on average 0.38 rem in addition to background radiation. People living just outside the 30-kilometre evacuation zone received about 0.5 rem in addition to background.

On anti-radiation drug studies and on radiotherapy research

The new National Centre of Radiation Medicine is a complex consisting of three institutes: for Clinical Radiology; for Experimental Radiology; and for Epidemiology and Radiation Injury Prevention. Professor V. Bebeshko, Director of the Clinical Radiology Institute is quoted as saying that by the end of 1987 the number of specialist and maintenance staff will exceed 600 and will eventually increase to 800 and that "the Centre has started looking for drugs that would prevent radioactive substances from accumulating in tissues and would help the body get rid of them" and that "the Centre will also do research unrelated to the effects of Chernobyl and that one subject of study will be X-ray therapy [radiotherapy], widely used in the treatment of malignant tumours [cancers]. The scientists will be looking for the best possible techniques in this area" (presumably using the most modern computer controlled radiotherapy linear accelerators: see caption to Fig. 136.)

FIG. 137. The turbine hall, May 1987.

Courtesy of TASS

FIG. 138. A panoramic view of all the units, May 1987.

Courtesy of TASS

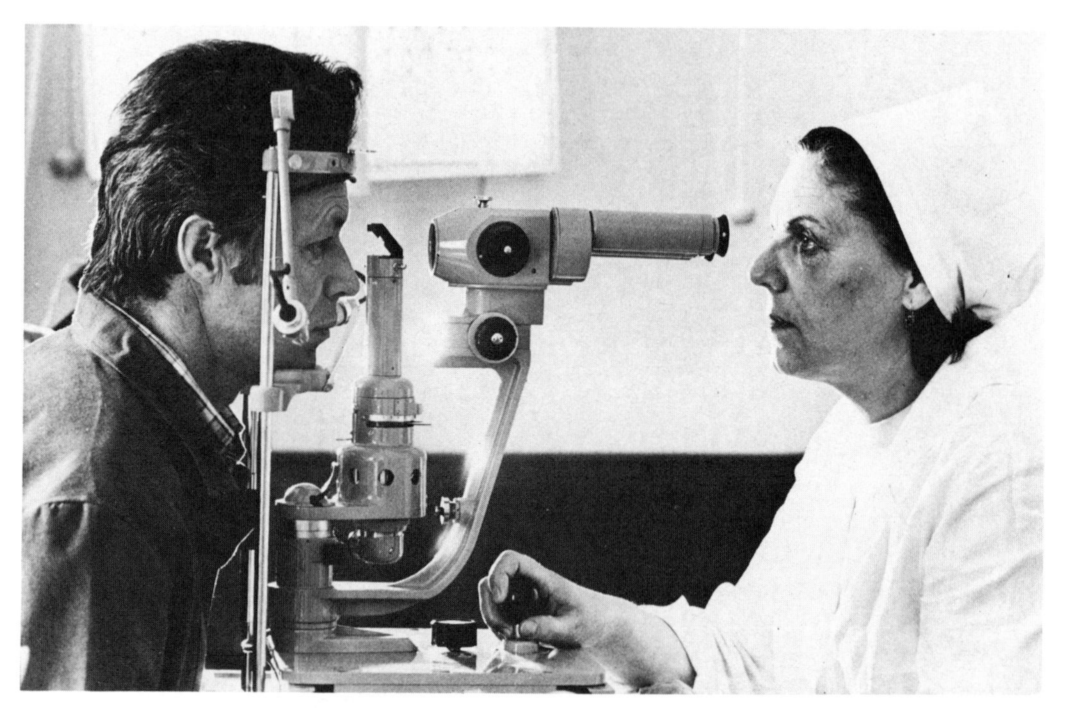

FIG. 139. Boris Shinkarenko, chief of the dosimeter instruments repair shop, undergoing a preventative medical examination at the oculist's, May 1987. All those involved in the Chernobyl accident and its aftermath of clean-up operations will be closely monitored medically. At present, there are not many medical follow-up reports, but in the *Independent* newspaper in the United Kingdom, 30 May 1987, Reuter's Moscow correspondent reported the following: "A Soviet film maker who worked in Chernobyl within days of the accident has died of radiation sickness, the weekly *Nedelya* reported. Vladimir Shevchenko, director of the film *Chernobyl: a chronicle of difficult weeks*, died two months ago, and two cameramen who worked with him were receiving hospital treatment. *Nedelya* said the film, which was screened at the Soviet film festival in Tbilisi last week, but has not been released in Moscow, had shocked its audiences." Another medical comment was reported in the UK newspaper *Today* on 8 June 1987 which stated that a magazine writer, Yuri Shcherbak, had interviewed, among others, a lone ambulanceman who was left to treat the first victims of the disaster and that: "He did not know what had happened inside the nuclear plant and did not at first recognise the symptoms of radiation sickness."

Courtesy of TASS

FIG. 140. Zelyony Mys, the housing settlement for shift workers at the Chernobyl plant. In May 1987 each shift lasted five days, followed by six days' leave in Kiev for operational personnel. For other personnel the shifts are fifteen days on and fifteen days off. Each worker at the plant now has to take practical and theoretical educational training, followed by compulsory examinations.

Courtesy of TASS

FIG. 141. Hothouse worker Zoya Maximenko and Vladislav Barsukovsky, the chief of a radiological research laboratory, monitor the flowering of vegetable marrows, May 1987.

Courtesy of TASS

Fig. 142. The sarcophagus, May 1987.
Courtesy of TASS

Суровые уроки Чернобыля

Завершился судебный процесс над виновниками аварии на АЭС

FIG. 143. These headlines from the 1 August 1987 issue of *Pravda* read: "Severe lessons of Chernobyl. Trial of culprits in the accident at the atomic energy station has finished." *Pravda* reported the end of the trial on the back page of the paper. The essentials of the article were as follows: The charges against the accused were a severe indictment against indiscipline and irresponsibility in the professional obligations of those involved as well as a serious lesson to us all. On 29 July the trial was completed of the criminal affairs of the former leadership of the station. It was under the chairmanship of Brize, a member of the Supreme Court of the USSR, with the participation of the Senior Assistant of the State Procurator of the USSR, Schadrin. The trial was held in Chernobyl and continued for more than three weeks. More than ten witnesses, as well as those who suffered, gave evidence, and this was analysed along with the conclusions of experts and specialists. All were given an opportunity to confirm once again the real reasons for the accident and to reconstruct its circumstances. It gave the opportunity to prove the guilt of those who were accused. The former director of the station, Bryukhanov, was considered the main culprit, having been director with a comprehensive responsibility. He was not reliable in carrying out regulations and safety instructions. Fomin, who was the former Chief Engineer, and his deputy, Djatlov, were also accused. Bryukhanov, Fomin, Djatlov, as well as former chief of reactor room, Kovalenko, did not have discussions as required since they did not analyse all the peculiarities of the forthcoming experiment and did not make the additional necessary measurements to provide safety. The former night shift chief, Rogoshkin, knew the situation but did not act as he should because he did not want to become involved (the literal translation from the Russian is that he "self escaped"). He did not monitor the experiment and when he received information about the accident he failed to activate the system for informing personnel. The former State Inspector of Gosatom & Energonadzor of the USSR, Laushkin, did not ensure that all the safety instructions and regulations were carried out at the atomic energy station. Bryukhanov, Fomin and Djatlov were given maximum sentence for such criminal offences, 10 years; Rogoshkin was sentenced to 5 years in prison, Kovalenko 3 years and Laushkin 2 years.

Courtesy of Pravda

FIG. 144. The judges in session at the court set up in Chernobyl at what was the House of Culture. The Chairman is Judge Raimond Brize of the Supreme Court. It was reported in the UK newspaper the *Guardian* on 30 July 1987 that in his summing up at the end of the trial he clearly stated that: "There was an atmosphere of lack of control and lack of responsibility at the plant" and that people had played cards and dominoes and written letters while at work. It also emerged from the outline of the case in the *Independent* of 8 July 1987, that other accidents had only narrowly been avoided at the plant, notably in 1982 and 1985, when elementary safety rules had been ignored, and supervisors had not been alerted.

Courtesy of Novosti

FIG. 145. The three main defendants at the trial, from left to right: Viktor Bryukhanov aged 51, Nikolai Fomin aged 50 and Anatoly Djatlov aged 57. No photographs were located of the other three defendants, Yuri Laushkin aged 50, Boris Rogoshkin aged 52 and Alexander Kovalenko aged 45. All except Laushkin were charged under Article 220 of the Ukrainian criminal code which covers violations of safety rules at plants where there is a danger of explosion. Laushkin was charged under Article 167 with negligent or unfaithful execution of his responsibilities. In addition, Bryukhanov was also charged under Article 165 which governs abuses of power. Bryukhanov and Fomin argued through their lawyers that the trial should be held in Kiev and not in Chernobyl, but they were unsuccessful in this plea. Djatlov was quoted (*Guardian*, 8 July 1987) as saying on the opening day of the trial: "With so many human deaths I can't say I'm completely innocent." He was at the plant during the accident whereas Bryukhanov arrived at 0200 hours, some 30 minutes after the explosion.

Courtesy of Novosti

FIG. 146. The former director of the power plant, Bryukhanov, during the trial.
Courtesy of Associated Press

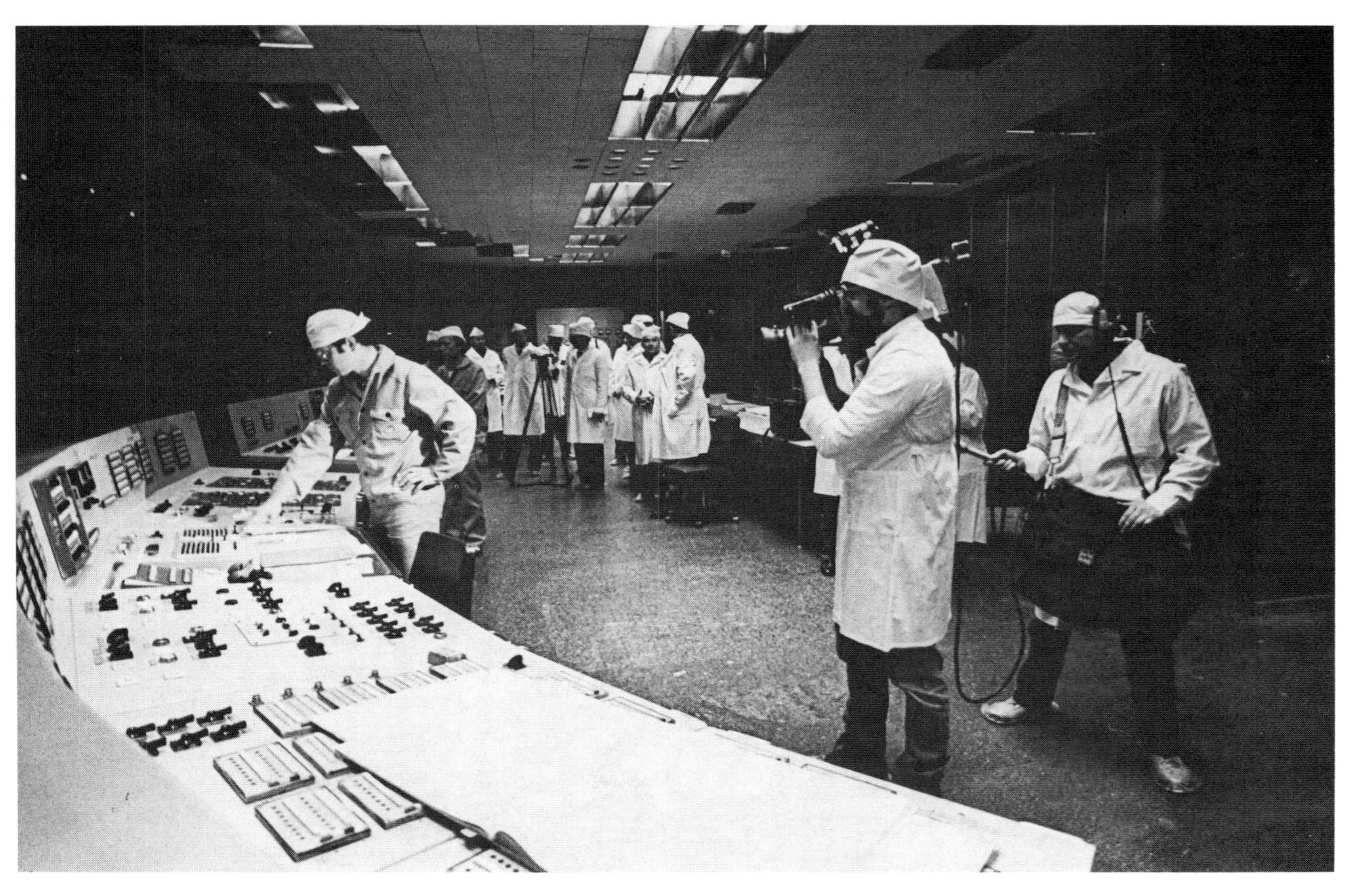

FIG. 147. Camera crewmen of the US Cable News Network filming inside the control room of the No. 1 reactor at Chernobyl as shown in the 5 September 1987 issue of *Soviet Weekly*.

Courtesy of Novosti

Fig. 148.

Courtesy of UNOST

FIG. 148. The engraving is from an article entitled "Chernobyl, Documentary Story" by Yuri Scherbak, which appeared in the June 1987 issue of the Soviet magazine *Unost*. It depicts a doctor, a Geiger counter operator, workers' flats in Pripyat, the roof of the reactor hall of unit No. 4 and a helicopter bombing mission. However, one of the most interesting features in the article, which is specifically aimed at a Soviet youth readership, is an eyewitness account by the first doctor to arrive at the scene of the accident, Dr Valentin Petrovich Belokon. It is a report which I have not found in any English language publication.

Dr Belokon was aged 28 at the time of the accident, with two young daughters, Tania aged 5 years and Katya aged 1½ months. He was also a sportsman who specialised in weight lifting, and was employed in Pripyat as an accident and emergency physician. His interview with Yuri Scherbak started: "On 25 April at 2000 hours I started my work in Pripyat, where there is an accident and emergency medical brigade consisting of one physician (myself) and a doctor's assistant (Sasha Ckachok) and six ambulances. On 25 April Sasha and I worked separately and my driver was Anatoly Gumarov. At 0135 26 April on my return to the medical centre I was told that there had been a call from the Nuclear Power Plant and that 2 or 3 minutes earlier Sasha had left for the NPP. At 0140, he telephoned to say that there was a fire, with several people burnt and that they needed a doctor. I left with my driver and arrived in 7 to 10 minutes. When we arrived, the guard asked 'Why don't you have special clothes.' I did not know it would be needed and was only wearing my doctor's uniform and since it was an April evening and the night was warm, I did not even wear a doctor's cap. I met Kibenok (a fireman lieutenant) and asked 'Are there patients with burns?' Kibenok's reply was that 'There are no patients with burns but the situation is not clear and my boys feel like vomiting.' My talk with Kibenok was near the energy block (unit No. 4) where the firemen stood."

Pravik (also a fireman lieutenant) and Kibenok had arrived in two cars and Pravik quickly jumped out of the car but did not come to see me. Kibenok was excited a little and alarmed. (Pravik and Kibenok were among the firemen who died, see Fig. 58).

What now follows is Dr Belokon's description of the first of the patients:

"Sasha Ckachok had already taken Shashenok from the NPP where he had been pulled out by workers after being burnt and crushed by a falling beam. He died on the morning of 26 April in a medical recovery room. The second patient was a young boy about 18 years old. He had vomiting and severe headache, and as I did not yet know about the high level of radiation I asked him what he had eaten, and how he had spent the previous evening. His blood pressure, at 140 or 150 over 90, was slightly higher than the normal 120 over 80 for an 18 year old. However, the boy was very nervous. At this time, the workers who came out of the NPP block were very disturbed and only exclaimed 'It is horrible' and that 'The instruments went off scale.' Three or four men from the technical staff all had the same symptoms of headache, swollen glands in the neck, dry throat, vomiting and nausea. They all received medication and were then put into a car and sent to Pripyat with my driver Gumarov. After that, several firemen were brought to me and they could not stand on their feet. They were sent to hospital."

Dr Belokon now begins to feel unwell and records that at 1800 on 26 April (many hours after his arrival): "I felt something wrong in my throat and had a headache. Did I understand it was dangerous? Was I afraid? Yes, I understood, Yes I was afraid, but when people see a man in a white uniform is near, it makes them quieter. I stood as all of us stood, without any breathing apparatus,* without any other means of protection.* When it became lighter (on 27 April) there was no fire to be seen in the block, but there was black smoke and black soot. The reactor was spitting, but not all the time, only as follows: smoke, smoke, then belch!! Gumarov arrived (back from Pripyat after taking the injured to hospital) and I felt weakness in my feet. I did not notice it when I walked, but now it has happened. Gumarov and I waited another 5 minutes to see if anyone else asked for assistance, but nobody did. That is why I said to the firemen 'I am going to hospital, if there is a need, call us again.' I went home, but before I washed and changed my clothes, I passed iodine to those in the hostel, asked them to close all windows and to keep the children inside. Then I was taken to the treatment department of a hospital by our Dr Diyakanov and given an intravenous infusion. I felt very bad and started to lose my memory, at first partially and then totally. Later, in Moscow, in Clinic No. 6 I was in one ward with a dosimetrist. He told me that just after the explosion, all instruments were off scale,* they called to the Safety Engineer and that engineer answered 'What is the panic? Where is the shift chief? When he is available tell him to call me, you yourself don't panic, such a report (about off scale measurements) is not correct.'"

Some time after this interview, in Autumn 1986, Yuri Scherbak met Dr Belokon in Kiev and as an after effect of the accident saw that Dr Belokon had breathing problems, where earlier, as a weight lifting athlete, he would certainly have had none. He now works as a paediatric surgeon in Donetsk. Scherbak ended his report by emphasising that this was the story of the first physician in the world who worked on site during the Chernobyl accident.

Courtesy of UNOST

---

* Statements also made in Vladimir Gubaryev's play *Sarcophagus*.

МЫСЛИ ВСЛУХ

На рентгене правды

Владимир Яворивский

FIG. 149. Under the headline "Censors pass damning exposé", *The Times* of 25 September 1987 described some of the contents of an article in the Russian magazine *New Times* which was written by Vladimir Yavorisky and entitled "The X-ray of Truth". It was contained in the 5 September issue of the magazine and the illustration shows the cyrillic title. The smaller lettering is a subtitle "Thinking aloud" with the main title literally translated as "On rontgen of truth". The connotation is presumably that just as an X-ray image shows the "truth" behind what was previously invisible, so does the Yavorisky article uncover previously unknown "truth". This information includes:

- the benefits for workers now at Chernobyl include an increase in annual holiday and, for rights related to years of service, one working day at Chernobyl is calculated as three working days.
- a local reporter from Pripyat, Lyubov Kovalevskaya, had published in *Literaturna Ukraina* prior warning of the possibility of a disaster, but had been ignored, and nearly suffered expulsion from Pripyat by Party officials.
- even when the plume of radioactivity was soaring into the sky over Pripyat, people trod on the radioactive debris because local Party chiefs did not give even elementary safety instructions and tried to cover up the scale of the accident.
- the director, Viktor Bryukhanov, informed the Kiev regional civil defence three-and-one-half hours after the accident that it was only a fire on the roof and that it would be put out.

Courtesy of Novoe Vremya

FIG. 150. A Chernobyl medal. This 6 centimetre diameter medal was presented by Soviet participants to Mr Meyer of the Division of Public Information, IAEA, on 6 October 1987, on the occasion of a meeting at the IAEA of Public Services International, a labour union group representing the nuclear power facility workers. Literal translation of the cyrillic lettering on the face of the medal is:

'TO PARTICIPANT OF LIQUIDATION OF THE ACCIDENT 1986
CHERNOBYL ATOMIC ENERGY STATION'

On the back of the medal the symbolism for peace is seen above the image of the power station.

Михаил Пантелеевич
УМАНЕЦ

Директор Чернобыльской АЭС

г. Припять
Телекс 132906 АТОМ Тел. 43359

Mikhail P. UMANETS

Director of Chernobyl Nuclear Power Plant

Pripyat
Telex 132906 ATOM Tel. 43359

Fig. 151. The business card, in Russian and in English, of the new Director of the Chernobyl nuclear power plant. There are essentially two organisations at present on site at Chernobyl: one for the operation of the power station, and one for the "clean-up" process. The logo refers to the latter and is termed "*KOMBINAT*" in cyrillic lettering, which in English would translate as Combine, presumably referring to a combination of services. "*KOMBINAT*" was formed on 2 October 1986.

Courtesy of Mr Adrian Collings, CEGB

Fig. 152. Mr Mikhail Umantes, appointed February 1987. Photograph taken April 1987 at a round table meeting "Chernobyl One Year After: Expert Opinion".

Courtesy of Novosti

FIG. 153. The administrative block of the Chernobyl nuclear power station, 12 October 1987. The long central banner is in place for the anniversary celebrations of the October 1917 revolution. Translated, it reads:

Towards the 70th Anniversary of the
October Revolution – Reliable and
Catastrophe Free Work!

The bottom left poster states:

We Say No to Nuclear Folly

and depicts a red fist shattering a black missile which has a white letter A at its tip to signify the nuclear warhead.[11] The two plaques on the doors on either side of the central clock pillar name the Chernobyl Nuclear Power Station, followed by the words "of V. I. Lenin".

The bottom right poster states:

The Acceleration of Scientific and Technical
Progress is the Key Question of the Economic
Policy of the Communist Party of the Soviet
Union (CPSU).

Courtesy of Mr Adrian Collings, CEGB

FIGS. 154, 155. Deserted Pripyat, 2 December 1987.
Photograph by R. F. Mould

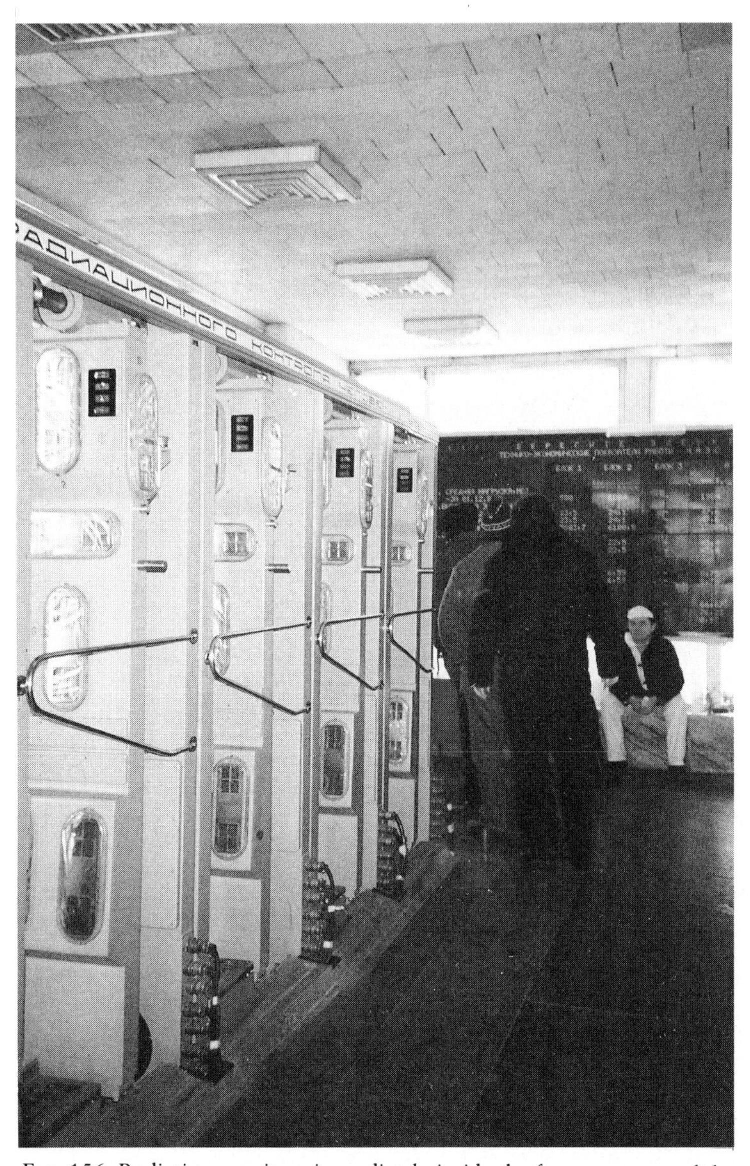

Fig. 156. Radiation monitors immediately inside the front entrance of the administrative building of the power station. The cyrillic lettering at the top of the array of monitors translates as Radiation Control. 2 December 1987.

Photograph by R. F. Mould

FIG. 157. View of the turbine hall from turbines Nos. 1 and 2 (linked to unit No. 1), looking towards the wall which now blocks off turbines Nos. 7 and 8 which were linked to unit No. 4. 2 December 1987.

Photograph by R. F. Mould

Looking toward Turbine No. 1.

Photograph by R. F. Mould

View on entering turbine hall.

Photograph by R. F. Mould

Part of control panel No. 1.

Photograph by R. F. Mould

FIG. 158. Mr Dmitri Chukseyev of the Novosti Press Agency measuring a dose rate of 3.9 millirem per hour, using a Soviet-made radiation monitor provided by Mr Alexander Kovalenko, Head of Information and Foreign Relations Department of AI "KOMBINAT". 2 December 1987.

Photograph by R. F. Mould

FIG. 159. The sarcophagus (to the left of the chimney). 2 December 1987.
Photograph by R. F. Mould

9

Follow-up

Professor Wilhelm Röntgen (1845–1923), the man who discovered X-rays in 1895, only gave one recorded interview—to Sir James Mackenzie-Davidson in 1896—and the following exchange was part of this interview:

MACKENZIE-DAVIDSON (*asking about the discovery*): "What did you think?"

RÖNTGEN: "I did not think, I investigated."

How illuminating it would be if we only knew how the surviving Chernobyl power plant staff who were on duty on 26 April 1986 replied to the question "What did you think (*you were doing*)?" which must surely have been asked during the follow-up to the accident. The start of their reply might well have been: "I did not think, . . ."

THE FUTURE is uncertain, both for civil nuclear power in some countries (but not I expect in the Soviet Union)* and for the late effects on the health of some of the more heavily irradiated members of the Soviet population who were in the 30-kilometre zone at the time of the accident. There is currently much debate among radiation medicine experts as to what, if any, are safe "low-level" doses of radiation, and this discussion will continue for years to come. Hard medical and statistical proof of cancer induction by low-level radiation doses is a virtual impossibility as there are so many other factors which would have to be taken into account and which could easily mask any "radiation effect".

The International Atomic Energy Agency

The International Atomic Energy Agency and its Director-General, Dr Hans Blix, are obviously pro-nuclear, and it is an inescapable fact that with the world's population rising steeply each year, there is an urgent need for energy sources. Alternatives to nuclear power such as solar energy will still take years to develop on a

* At the 22 April 1987 news conference in Moscow (reported by TASS) on the effects of the accident, the Soviet Minister of Atomic Energy, Nikolai Lukonin, presented future plans to double the generation of electricity at nuclear power stations by 1990 as compared with 1985 and to more than treble it by 1995. Eleven new nuclear power stations are now under construction in the Soviet Union, mainly in the European part of the country.

large scale, and with the present coal-fired power stations, acid rain pollution is a major drawback, although costly measures could be taken to prevent this. Any rundown of a nuclear power industry has, in practice, to take place over a period of years, with the alternative energy sources being phased in during the same period. So, like it or not, civil nuclear power remains the only viable alternative for the foreseeable future, and the recommendations of the IAEA following Chernobyl should be taken seriously.

In particular, the safety of existing nuclear power plants, especially the Soviet RBMK1000 and RBMK1500 plants, should be thoroughly reviewed so that "Chernobyl" cannot be repeated at any of the other Soviet plant sites. The training of nuclear power plant staff is absolutely essential. The sequence of events in Chapter 3 have shown how incredibly lax were the Soviet operators on duty on 25–26 April 1986. If this is typical in the Soviet Union, then what might the future hold if operator training is not improved very significantly?

Most countries with nuclear power have been claiming how safe they are, as their designs of reactor are different from the RBMK, and indeed this is probably true in many cases; but there should be no complacency. A list of nuclear accidents, including Windscale in 1957 and Three Mile Island in 1979, are frequently being reiterated in the world's press. Indeed, they should not be forgotten. However, the horrendous list of operator mistakes at Chernobyl, coupled with some design faults in the RBMK reactor, make the events of 26 April 1986 something the world cannot afford to repeat. On a personal level, you only have to read Chapter 4 and the case histories of the heroic firemen, whose working conditions immediately after the accident are hard for us to imagine: 30 metres above the ground a roof where the bitumen had melted in the heat, so that their boots were leaden; soot and smoke made it difficult to breathe; their clothing was not designed for protecting them in such conditions; and it was impossible to extinguish some of the fires using water and chemicals because these would just evaporate. Later, in spite of excellent medical care and attention, a large number of the radiation-injured firemen also were subject to viral infections in their radiation damage wounds. These are facts.

However, the 25–29 August 1986 meeting in Vienna adopted a number of proposals for expanding and strengthening the Agency's activities in the field of nuclear safety. They included suggestions that the international programme on experimental and analytical research on severe accident sequences should be expanded; and that international exchange of information on the man-machine interface (this interface seemed to be non-existent on 26 April at Chernobyl) should be further promoted, as should the exchange of experience on operating and training methods. In all, the list contained thirteen points for areas of future collaboration under the aegis of the IAEA. These were:

1. Severe accident sequences.
2. The man–machine interface.
3. The balance between automation and direct intervention by an operator.
4. Exchange of experience in operator training procedures and management, with the possibility that the IAEA will consider international accreditation of operators.
5. International safety standards to be reviewed.

6. Fire protection standards to be upgraded.
7. International emergency reference levels to be set.
8. Decontamination.
9. Radioactive dispersion in the environment: air, water, food chain.
10. Assessment of individual and collective radiation doses.
11. Optimisation of epidemiological methods.
12. Efficiency of treatment procedures for radiation sickness and radiation burns.
13. Efficiency of treatment procedures for late effects on health.

Visits to Chernobyl: December 1986–October 1987[12]

Three and one-half months (16–19 December 1986) after these recommendations were made, the first visit to Chernobyl by a politician from a Western country was permitted. This was by **Mr Peter Walker, the Energy Secretary in the British Government**. The joint Soviet Union–United Kingdom communiqué for this visit read as follows:

"In accordance with the final document of the special session of the general conference of the IAEA held in September, both sides recognise that nuclear power will continue to be an important source of energy for social and economic development and emphasised that the highest level of nuclear safety will continue to be essential to the use of this energy source. Noting that the final document urged strengthening of international co-operation, at both bilateral and multilateral level, both sides have agreed to strengthen bilateral contacts in the following areas:

Questions of safety at nuclear power stations including automatic control systems.
Radiological protection, including decontamination techniques.
Waste disposal techniques.

Bilateral discussions on these topics will take place in 1987.
Both sides confirmed their intention to assist the IAEA in its work of implementing the recent international conventions on early notification and mutual assistance in the event of nuclear accidents, which both countries have signed. They reaffirmed the central role of the IAEA, under its statute, in encouraging and facilitating international co-operation in the peaceful use of nuclear energy, including nuclear safety and radiological protection."

In addition, it was reported in the 18 December 1986 issue of *The Times*, that a team of British doctors and medical experts is to go to Kiev in 1987 to study health problems caused by the accident. In practice, this was only one suggestion among many that were raised by Mr Peter Walker with his Soviet Energy Minister counterpart. If such a visit were to take place, as well it might, it would be arranged between the Ministries of Health and Foreign Offices of the Soviet and United Kingdom governments. (*By 2 December 1987 it had not taken place.*)

Very few visits are being allowed to the Chernobyl power plant site by politicians, journalists, trade unionists or scientists from Western countries. Apart from Peter Walker's visit, a 6-day visit to the Soviet Union was also made by John Edmonds, the

United Kingdom General Secretary of the General, Municipal, Boilermakers and Allied Trades Union, who was accompanied by four colleagues. This particular visit was reported by John Edmonds in *The Independent* newspaper of 4 December 1986, and includes a comment by Stephan Shalnyev, chairman of the Soviet Trade Union Centre: "We had become very complacent, Chernobyl punished us for it." There is also mention of Nikolai Simochatov, president of the Soviet power workers union, who commented that "Even experienced people do stupid things." He was actually at the power plant hours after the explosion and stayed for a 1-month period with only a single short break. Edmonds describes Simochatov's current state of health as "Complexion pallid, unable to eat anything cold, and has repeated bouts of a flu-type illness." However, the trade union group were not permitted to visit Chernobyl itself, and undertook their discussions in Moscow, Kiev and in Voronezh, the site of a power station on the River Don. During their initial discussions the obvious question of how engineers and operators could violate so many rules was raised. This was met with a shrug of the shoulders, a rueful expression and the counter-question "Why does an experienced driver cross the traffic lights at red?" This group was also informed that the Chernobyl deputy engineer "was in gaol, for his own protection", and it then transpired that the reason was "to stop him committing suicide". This indicates the deep psychological shock of at least one senior member of the Chernobyl power plant staff.

The first visit by Western journalists to the Soviet Union to report on the effects of Chernobyl was announced by TASS on 12 December 1986, and specifically mentioned were the *Washington Post*, *New York Times*, *Chicago Tribune*, the London *Times* and the American TV network, NBC. They travelled in the Ukraine and saw the new villages of Zdvizhevka (which is 60 kilometres from the plant) and Nebrat and spoke to evacuees, Chernobyl plant workers, firefighters, bus drivers, physicians and officials. This visit, though, did not appear to include the Chernobyl nuclear power plant itself. More recently, however, cameramen from the **US Cable News Network** have been allowed to film inside the control room of the No. 1 reactor (see Fig. 147).

The Times of 16 December, reporting on the earlier visit by the press, spoke of evacuees quite naturally wishing to return home and quoted as a typical interview example "Anastasia Panasivona, aged 72, a peasant, who complained that with her new central heating, she missed being able to sleep on her stove, a favoured Ukrainian custom for combating the rigours of winter". These winter rigours in mid-January 1987 must have been very noticeable. Parts of the Ukraine were some 2–3 metres deep in snow, and the Soviet Union experienced its coldest winter for 50 years. For instance, on 10 January the temperature dipped to −39° Centigrade (70° of frost on the Fahrenheit scale). This was very close to the record low this century of −42.2° Centigrade set on 17 January 1940. In northern Siberia, it was even colder, being 60° below.

The first visit to Chernobyl by Dr Hans Blix and Mr Morris Rosen of the IAEA was on 5–9 May 1986, and a second visit took place from 11–16 January 1987. This was followed by press briefings in Moscow and at Vienna Airport. The essentials of the Vienna briefing were:

—Unit No. 1 was visited and it was reported that there was no contamination of the facility and that the unit could be entered without wearing protective clothing.

—In the 10-kilometre zone around the plant, there would be no resettlement of people for the foreseeable future, except for a special village for the power plant operators.

—In the 10-kilometre to 30-kilometre zone there is at present some resettlement in twenty villages.

—When asked about the attitude of the population to resettlement in the original villages, the reply was that the old people were happy to be returning, but that the young people were content to remain in the new villages where they were housed after the evacuation.

Additionally, TASS reported that it was agreed that "the IAEA would arrange a meeting of experts in the first half of 1987 to discuss appropriate methods to study possible long-term consequences of radiation's effect on people". Some input to the meeting will no doubt arise from the 11 January 1987 visit of a delegation of Soviet physicians to Japan in response to an invitation from Hiroshima doctors to study the experience and methods of treatment of the "Hibakushas", the victims of Hiroshima and Nagasaki.

It is also of interest to note that by the time of the second IAEA visit, Chernobyl nuclear power plant units Nos. 1 and 2, with their total of four 500 megawatt turbines, had been in operation for some time and since switch-on after the accident had generated more than 1600 million kilowatt-hours of electrical energy. Prior to the accident the energy generation by the entire Chernobyl power station was more than 100 billion kilowatt-hours of electricity.

To me, the most unusual report of a visit to the Soviet Union relating to Chernobyl events was issued by TASS in two short reports on 17 December 1987 and I reproduce the first report in full.

"Prominent American author **Frederick Pohl, Vice-President of the World Science Fiction Society**, spent 14 days in the USSR collecting material for his new book, the newspaper *Literaturnaya Gazeta* said today. Pohl said that this book would be devoted to Chernobyl events, it is not a documentary story, but a piece of fiction with fictitious characters, whose number is more than forty. Pohl familiarised himself with all the circumstances of the accident at the nuclear power plant and talked with eye-witnesses. Pohl considers his aim to be achieved and talked with people who participated in eliminating the consequences of the accident, and those who were fighting flames in the very first hours. Even before his arrival he knew some details. Here he discovered the most important aspect of the problem, as he believes this will help him tell the American readers the truth about Chernobyl."

The second report is concerned with comments about press distortion and the Chernobyl victims' fund, but also includes the following:

"My book is going to be neither anti- nor pro-Soviet, since in the novel, like Chernobyl, there are instances of heroism, selflessness and courage; but there was also cowardice, as there were people who fled saving their own selves."

Books can either be fact, fiction or a mixture of fact and fiction (sometimes termed faction). However, I cannot see how a fictional book about Chernobyl will help American readers to understand the accident and its aftermath. However, it could add confusion, since no reader will be able to distinguish the factual parts from the fictional parts. Was this a useful visit?

The visit of the General, Municipal, Boilermakers and Allied Trades Union has already been referred to, and on 2–5 April 1987 this was followed up by a delegation visit of the **Nuclear Energy Review Body of the Trades Union Council of the United Kingdom.** This delegation not only held meetings with the Soviet Minister of Atomic Energy, the USSR Academy of Medical Sciences and the Ukrainian Council of Ministers, but also actually visited the Chernobyl site, reporting their tour of the control room and turbine hall as follows:

"Toured the control room of unit No. 1 and the turbine hall and saw the turbine generator sets associated with units Nos. 1 and 2. Unit 1 was running, unit No. 2 was shut down for planned maintenance. The generators for units Nos. 3 and 4 were concealed by specially constructed walls. They did not visit the entombment of unit No. 4, because this had not been allowed for in the programme."

Other items of interest in this TUC delegation report include the following.

On radiation doses

The Soviet All Union Central Council of Trades Unions "resolutely supported further tightening of safety and hygiene practices in nuclear energy. Present practices were based on international standards, including the 5 rem (50 mSv) per person per year limit for nuclear workers. Soviet nuclear stations normally achieved lower dose levels than that, usually about 1 rem (10 mSv) per year and at the Kursk plant it is as low as 0.18 rem (1.8 mSv). The AUCCTU believed that the existing international standards were too high and that they should be reduced."

Academician Ilyin reported that some time ago the Director of Nuclear Safety in the Soviet Union specified a maximum dose level of 25 rem (250 mSv) per annum in accident situations and that during the Chernobyl accident this dose was received by some personnel within minutes. He also reported that "the majority of doses which had been received by the population most affected by the accident were in the range 1.5 to 10 rem (15 to 100 mSv)".

During the visit to the Chernobyl site, the Station Manager, Mr Mikhail Umanets, discussed the decontamination of units Nos. 1 and 2 and stated that monitoring records for these two units, for the first 3 months after decontamination, showed "all staff were working under 4.5 rem (45 mSv) per annum".

On induction periods for radiation-induced cancer in man

Academician Ilyin reminded the TUC delegation that "cancer as a result of radiation occurs many years after exposure".*

* I have reviewed existing data on induction periods in my book on *Cancer Statistics* and refer to two publications in 1957. In one, for twenty patients treated with X-rays for benign (i.e. not cancer) conditions who later developed skin cancer, the induction period was in the range 14–45 years. In the second, for thirty-four case histories, the range was 13–44 years for nine cases of basal cell carcinoma, 12–56 years for eleven cases of squamous cell carcinoma, and 10–55 years for the six cases of sarcoma. For the entire series of thirty-four patients the mean time was 30.1 years, the minimum was 8 years and the maximum was 56 years. Populations exposed to medical irradiation in the early years of this century (and those exposed to X-rays for cosmetic purposes such as epilation "beauty treatments") should not be equated to the atomic bomb populations of Nagasaki and Hiroshima when considering

On milk consumption

At the Academy of Medical Sciences meeting it was said that by 30 April Soviet scientists had agreed a contamination level in milk of up to 3700 becquerels/litre, in particular for iodine-131, and that some 46,000 had been given iodine administrations by midnight on 26 April.

On the health of the Chernobyl firemen

Some twelve to thirteen firemen were still suffering from the after-effects of the accidents to varying degrees.

On evacuation measures

At the meeting with the Council of Ministers of the Ukraine, the Chairman, Mr A. P. Lyashko, gave the following statistics. By 1 September 1986, 8210 houses had been built, 7500 flats in Kiev had been allocated to evacuees, as had 2000 hostels for young people. A further 4500 houses were built by 1 January 1987. To accommodate the Chernobyl workers, a new village, called Slavutich (this is the old name of the River Dnieper), was being built to house 6000 people. It would be completed by June 1988. 500,000 mothers and children had been sent to Black Sea resorts and 300,000 had been taken to pioneer camps.

On decontamination

Mr A. P. Lyashko stated that "road surfaces and 5 cm of top soil throughout the entire 30-kilometre exclusion zone had had to be lifted and removed to remote areas".

On prospective additional cancer deaths

Mr A. E. Romanenko, the Director of the special Institute of Radiological Medicine established in Kiev after the accident, gave between 200 and 600 as his best current estimate of the number of additional cancer deaths resulting from the accident, and said that he felt that this estimate could be higher, but would not exceed 1000. The TUC delegation pressed Mr Romanenko for further details but his reply was "these [figures] were unofficial and that it was not possible at present to put these in writing, but it was intended to publish them in 1988".

On health monitoring

Mr A. E. Romanenko stated that the Radiological Institute intended to monitor for three generations and that systematic monitoring was already underway for 500,000 people in the Ukraine and 1,050,000 people throughout the Soviet Union as a whole.

---

induction of cancer. If comparisons are to be made, then the Chernobyl experience in future years ought to be compared with the medical irradiations pre-1920 rather than with atomic bomb experience. It is just not true, as some of the media might have us believe, that an epidemic of cancer in the Chernobyl population will be seen within some 2 or 3 years after the accident.

On land cultivation

At the meeting with the Council of Ministers of the Ukraine, land cultivation statistics were given as follows. "Around 50,000 square kilometres had been taken out of cultivation, fenced off with barbed wire and patrolled by security personnel. The uneven spread of contamination has meant that the furthest contaminated patch was 15 kilometres beyond the 30-kilometre exclusion zone. It was estimated that it would take 5–7 years to bring back all the land into cultivation, and the aim was to do so at a rate of 15% to 20% each year."

On the health of mothers, children and babies

The TUC delegation visited the Kiev Research Institute of Paediatrics, Obstetrics and Gynaecology and the Centre of Mother and Child Care of the Health Ministry of the Ukraine. The Director, Dr E. M. Lukyanova, gave the following statistics. "The Institute had dealt with 2600 pregnant women, 1160 babies, 3100 teenagers and 500 other women of child bearing age. Measurements were taken of the accumulation of iodine-131 from May–July 1986 and this had given a dose to the thyroid of not more than 10–13 rad. New babies had been examined on the second or third day after birth but no significant iodine-131 influence had been detected. By the end of July 1986, work was concentrated on identifying radioactive isotopes other than iodine-131, which could have been accumulated. The 2600 pregnant women tested had shown caesium-137 concentrations not in excess of 0.01–0.30 microcuries. The level of radionuclides in mother's milk did not exceed permissible levels and it was quite acceptable for women to breast feed their babies. Tests had also been carried out on the water surrounding the foetus in pregnant women, and this did not show any significant radioactive concentrations. Among newborn babies the concentration had never exceeded 0.01 microcuries. The number of abortions and miscarriages had not risen at all."

Another visit to the Chernobyl area is reported in *The Times* (10 June 1987, "An eerie silence in the shadow of Chernobyl", 11 June 1987, "Chernobyl: where no money can change hands"). It describes a visit early in June 1987 by "**a handful of Western reporters**" and includes the following:

"The continuing dangers were quickly brought home when we were required to sign special forms and warned never to open bus windows or smoke in the 18-mile exclusion zone surrounding the plant. We were also ordered never to step on to roadside verges, nor to drink water, and we had to wear dark glasses if the sun shines brightly."

Mr Alexander Kovalenko,* a scientist who operated a radiation monitor at the time, measured ground radioactivity levels in the town of Chernobyl and at the plant, and the values in millirontgens were stated by *The Times* to be 0.1 and 12.8 respectively, (millirem/hour). In Kiev, 90 miles distant from the plant, the measurement was 0.02 (millirem/hour). Another lasting impression occurred when the reporters' bus came around a bend in the road to the Chernobyl nuclear power plant

* This is the Head of Information & Foreign Relations Department, AI "KOMBINAT," whom I met on 2 December 1987.

and "suddenly came across a surrealistic dump of more than 2000 abandoned cars and motorcycles".

The first news of the forthcoming trial of the power plant's senior personnel at the time of the accident was also reported on 10 June, and attributed to the scientist Mr Kovalenko.

"On 5 July the plant's former Director, Mr Viktor Bryukhanov, the former chief engineer and his former deputy would be brought back into the zone from a Kiev gaol to face trial on charges of criminal negligence. The trial will take place under a judge from the Soviet Supreme Court and will be held in Chernobyl's former House of Culture. Two of the defendants have already argued unsuccessfully that they have already suffered high radiation doses and should not have to return to the zone."

On 11 June *The Times* reported that they had been told by a senior Communist Party official that only 300 of the 92,000 refugees from Ukrainian territory (there were others from Byelorussia) had so far been permitted to return.

The headline about money not changing hands refers to the settlement of Zelyony Mys, 27 miles from the plant, which is used for workers and where no money is used. However, since the workers have everything provided for them, it is probably for this reason that the comment was made and not necessarily because of radioactive contamination. However, everyone at Zelyony Mys is regularly monitored from head to toe with radiation monitors and "shoes have to be dipped into a long tin bath".

A later *Times* report, 18 June, referred not to the visit but to a recent edition of the Soviet magazine *Argumenti i Fakti* which contains an article by Professor V. Knizhnikov of the Soviet Health Ministry. He described "radiation phobia" following the Chernobyl accident as "a fear which is first and foremost the result of lack of objective information and poor training of doctors in radiation medicine" and added, "but this has meant that, in some places, women had dangerous abortions late on in their pregnancies. Parents were afraid of giving their children milk, believing that it was contaminated, as a result of which cases of rickets have been registered." Professor Knizhnikov also disagreed with the estimates of Dr Gale (which are given later in this chapter) that in the next few decades 75,000 would die from cancer resulting from the disaster.

Very little else has appeared in the media in May and June, concerning follow-up of the accident, except for a small item in the 30 May 1987 issue of the *Independent*, which quotes the Soviet weekly *Nedelya* reporting the death of a Soviet film maker who worked in Chernobyl within days of the accident. "Vladimir Shevchenko, director of the film 'Chernobyl: a Chronicle of Difficult Times', died 2 months ago of radiation sickness, and two cameramen who worked with him were receiving hospital treatment."

The Central Electricity Generating Board in the United Kingdom have a well established technical exchange programme with their Soviet counterparts. As part of this programme, three CEGB personnel visited the Soviet Union for a week in October 1987 and this was scheduled to include a 1-day visit on 12 October to the Chernobyl nuclear power plant site. In addition, they also visited Zelyony Mys. Items of interest during the visit included the following:

On radiation doses

All three CEGB visitors wore film badges and one carried a pocket pen radiation dosimeter and this indicates an openness on the part of the Soviet authorities to allow visitors to make their own personal radiation monitoring measurements. The measurements using the pocket pen dosimeter were 11 microsieverts (μSv) during the flight from London to Moscow and 31 microsieverts (μSv) during the 8–9 hours on 12 October when they were within the 30-kilometre zone. The time spent actually in the nuclear power plant and the surrounding area was from 1100–1800 hours and during this period one of the Chernobyl staff measured a dose rate of 5 millirem per hour rising to 40 millirem per hour on the ground, at a distance just less than 200 yards from unit No. 4. It was stated that just after the accident the dose rate at this same point was several thousands of rem per hour.

On forests and soil removal

The area around the Chernobyl power plant was previously densely forested (as seen from Fig. 74), but by 12 October most of these trees, which were in an area of 2 kilometres × 1 kilometre, had all been cut down and buried. The earth removal in the vicinity of the power plant was said to be to a depth of some 1.0–1.5 metres.

On units Nos. 3, 5 and 6

It is planned to put unit No. 3 back into operation during November/December 1987. The building work before the accident was already quite well advanced for units Nos. 5 and 6, and particularly for No. 5. The cranes are still in position around these units and it is planned that in the year 1990 a decision will be made whether or not to continue with the construction.

On the towns of Pripyat, Chernobyl and Slavutich

By October 1987 one-quarter of Pripyat had been fully decontaminated. Its long-term future is that it is unlikely to be a dormitory town for power plant workers ever again. This is not because of decontamination problems, but because workers will by then have been housed in better accommodation than that of Pripyat (e.g. Slavutich). Also when building a town it is necessary to first provide the necessary infrastructure such as medical services, schools and the transport system. Repopulation of Pripyat does not therefore appear to be a worthwhile exercise. A possible alternative use in the long term would be as a scientific/research centre[13] which will undertake work relating to plants, trees and birds. It is not known why, but at the time of the accident the bird population deserted Pripyat and only returned in the summer of 1987. In Chernobyl, only 10 kilometres from the power station site, an administrative block of buildings has been built for staff, but very few people are actually living in Chernobyl. Those workers who do live in this town operate a shift system of 2 weeks on and 2 weeks off duty. In the new town of Slavutich, each of the fifteen Soviet republics will be responsible for building part of the town, and each will build in its own typical architectural style.

On Chernobyl nuclear power plant workers

There are still soldiers involved in operations around the power plant complex. The labour force, particularly that for construction work, is in part formed from volunteers from all parts of the Soviet Union. They have free accommodation and food, and work 8–10 hour shifts with 2 weeks on duty followed by 2 weeks off duty. Their wages are between two and four times standard wages.

On evacuation of Pripyat following the accident

Pripyat was evacuated in 36 hours and there has been some criticism of the time taken. It was pointed out to the CEGB visitors that there was only a single road out of Pripyat and that this passed the nuclear power plant particularly close to unit No. 4 which was the end unit nearest to Pripyat. It would therefore have been unacceptable, for example, to suggest that the population walked away from Pripyat along this road, which led right through the most highly contaminated area. Not enough private cars would have been available and therefore it was essential to provide adequate numbers of buses.

Health

Health effects following Chernobyl are of great public concern, but the figures of excess cancer deaths due "solely" to the accident varied widely in the media. Thus:

Daily Telegraph newspaper in the United Kingdom, 26 August 1986

Under the headline "Chernobyl cancer will kill 48,000" was a statement that the mortality could reach 48,000 and not the 6500 originally estimated. Later in the article it gave a figure of "200,000 cancer fatalities in the USSR".

London Evening Standard, 26 August 1986

Referring to the IAEA meeting then in progress, the *Standard* reported that the Soviet Union estimated that around 6000 people will die, and then went on to say that other delegations put the "cancer death toll as high as 50,000".

Wall Street Journal, 29 August 1986

A summary gave the following: "according to the Russians 6500; 45,000 according to the Natural Resources Defense Council (in the United States); and 24,000 according to experts at the IAEA". The newspaper did, however, have the sense to add that "all of these figures, of course, are merely statistical extrapolations" and they also drew attention to the natural disaster that had recently occurred, a volcanic eruption in Cameroon, in which some 1500 people were poisoned overnight.

The Times, 23 August 1986

"The Russians estimate that an extra 6530 cancer deaths may eventually result from the accident over periods of up to 70 years."*

* In many reports the figure of the 70-year time period was often omitted, implying that the deaths would all occur "soon".

New Scientist, 14 August 1986

The estimates of two Americans, von Hippel and Cochran, were reported by the *New Scientist*:

—"Tens or even hundreds of thousands of tumours and possibly several thousand deaths from cancer during the next 30 years.

—12,000–40,000 cases of thyroid tumours from iodine-131 inhalation. Only a few percent of these will be fatal.

—10,000–25,000 cases of potential thyroid tumours from iodine-131 absorbed from contaminated milk.

—8500–70,000 cases of cancer from all sources of caesium-137. About half might be fatal."

New Scientist, 11 September 1986

The highest estimate of mortality yet was given at a recent meeting of the American Chemical Society by Gofman: "More than one million people exposed to fallout from Chernobyl would develop cancer. About 500,000 of these cancers would be fatal."

One of the most recent estimates of predicted excess cancer deaths was given by Dr Robert Gale at the 11 April 1987 seminar of the Institute of Biology in London. His predicted consequences of the Chernobyl accident are given in the table below and include not only cancer, but also severe mental retardation and genetic damage. Any

		Cancer		
Population	Group	Excess cancer	"Spontaneous" cancer	Gale predicted per cent increase
Chernobyl	135,000	0–400	17,000	2%
USSR	280 million	0–20,000	27 million	0.07%
Europe	400 million	0–30,000	72 million	0.04%
Northern Hemisphere	3500 million	0–75,000	600 million	0.01%
		Severe Mental Retardation		
		Excess retardation	"Spontaneous" retardation	
Chernobyl	135,000	0–20	13	140%
		Severe Genetic Abnormality in 1st Generation		
		Excess abnormality	"Spontaneous" abnormality	
Chernobyl	135,000	0–100	7000	1.5%

excess cancer deaths at all represent a bad situation, but the *absolute numbers* (e.g. 0–400) cannot be viewed in isolation to those which would normally be expected (e.g. 17,000). These were termed by Gale as "spontaneous", but this should not be taken to mean "without any known cause", since they include the "self-inflicted" smoking-related lung cancers. *Absolute numbers and percentage increases* (e.g. 2%) place the cancer consequences in perspective. At the same seminar Mr T. Hugosson predicted a total excess of 100–200 cancer deaths in the next 50 years in Sweden.

However, not many people seem to realise that it was only during 25–29 August at the IAEA meeting in Vienna that any real radiation dosimetric data became available from the Soviet Union and included in the delegation's working documents. It would be a major research project to estimate numbers of cancer deaths over a period of 70 years (for instance, the age–sex structure of the population in this period must be predicted) from low-level radiation doses. Newsmen from Europe and the United States were clamouring all over the delegates for instant estimates—and some were foolish enough to give them!

In 1988 a report will be issued by UNSCEAR on the assessment of the long-term impact of the accident and we will then hopefully be a little wiser about the future risks to health. It will no doubt include for the various affected countries estimates of collective dose equivalents for different pathways such as cloud gamma radiation, inhalation, deposited gamma-radiation and ingestion; and thyroid and whole body measurement results. However, till that time the best one can do is to read the 28 August 1986 IAEA press conference report on "Health Effects", which was delivered by Dr Dan Beninson, that day's chairman of the working group on the topic, and the INSAG statement on "Late Stochastic Health Effects" which was published in December 1986 in the IAEA safety series summary report of the post-accident review meeting.

From the IAEA report

"Published estimates of the potential number of so-called 'excess cancer deaths' which may result from exposure to radioactive materials released from the Chernobyl reactor during the next 70 years may be ten times too high. Calculations in the technical report on the accident presented to a post-accident review meeting this week were deliberately conservative, and had been made on a pessimistic theoretical basis rather than by using actual measurements. Dr Dan Beninson, chairman of the International Commission on Radiological Protection and Director for Licensing of Nuclear Installations in Argentina, recalled that the media had been reporting expectations of as many as 20,000 deaths from malignant tumours in excess of those which would occur naturally among the population which had been exposed to radiation after the accident. 'That is nonsense, in many respects', he said. Dr Beninson explained that participants in the meeting had now presented data which suggested that a more accurate calculation would yield an expectation ten times smaller. Instead of 20,000 excess deaths, there may be only 2000. That does not mean that if we have only 2000 instead of 20,000 it is good, or more acceptable. It is a very bad number in any case, but it has to be put into perspective. That perspective could be acquired by looking at the doses actually expected to be received by people in the exposed population as a result of the accident, which were about 4.5 millirem/year. Naturally occurring background radiation accounted for an exposure of about 200 millirem/year. I can assure you that by going up a mountain and staying there you would get a larger dose than this. Going from my country (Argentina) to La Paz in Bolivia would multiply by two the cosmic ray dose you would receive, because La Paz is higher."

From the INSAG statement

"The magnitude of the health impact of the late stochastic effects, mostly neoplasmic and genetic in nature, can be assessed only after evaluation of the resulting collective doses. The information in this regard from the Soviet Union is preliminary and tentative. From the information available it appears that over the next 70 years, among the 135,000 evacuees, the spontaneous incidence of all cancers would not be likely to be increased by more than about 0.6%. The corresponding figure for the remaining population in most regions of the European part of the Soviet Union is not expected to exceed 0.15% but is likely to be lower, of the order of 0.03%. The relative increase in the mortality due to thyroid cancer could reach 1%.

"The number of cases of impairment of health due to genetic effects may be judged not to exceed 20–40% of the excess cancer cases. There is no information at present on which possible consequences of the *in utero* irradiation of human foetuses within the 30-kilometre zone could be assessed. Data on collective doses from other countries and in the process of evaluation and the assessment of possible stochastic consequences must be deferred to a time when these data become available. Preliminary estimates have been made of the doses to individual members of the public in the Soviet Union and of the dose to the population as a whole. These estimates will be refined as more data become available, and the overall radiological consequences of the accident will be assessed by UNSCEAR, in co-operation with the IAEA and WHO, on the basis of data collected from member states. International discussions should be held about the methodology for an epidemiological study of workers and selected groups of the population in the region of the plant."

APPENDIX 1

USSR State Committee on the Utilisation of Atomic Energy: The Accident at the Chernobyl Nuclear Power Plant and its Consequences

Information compiled for the IAEA Experts Meeting
25–29 August 1986, Vienna
Working Document for the Post-Accident Review Meeting

LIST OF CONTENTS

[7] Monitoring of environmental radioactive contamination and health of the population.
[8] Recommendations for improving nuclear power safety.
[9] The development of nuclear power in the Soviet Union.

Part II. Annexes 1, 3, 4, 5, 6

[1] Water-graphite channel reactors and operating experience with RBMK reactors. (5 pages)
[3] Elimination of the consequences of the accident and decontamination. (5 pages)
[4] Estimate of the amount, composition and dynamics of the discharge of radioactive substances from the damaged reactor. (20 pages)
[5] Atmospheric transport and radioactive contamination of the atmosphere and of the ground. (16 pages)
[6] Expert evaluation and prediction of the radiological state of the environment in the area of the radiation plume from the Chernobyl nuclear power station (aquatic ecosystems). (8 pages)

Part II. Annexes 2, 7

[2] Design of the reactor plant. (186 pages)
[7] Medical-biological problems. (70 pages)

Note

The International Atomic Energy Agency has published a report on the post-accident review meeting, which was written by INSAG (the International Nuclear Safety Advisory Group) and issued in October 1986 in the IAEA Safety Series (No. 75-INSAG-1) of publications. This INSAG report was requested by the Director General of the IAEA and states that "the main body of the INSAG report is derived from the excellent reports presented by the Soviet experts at the meeting" and that as "this material (i.e. the Soviet reports) can hardly be condensed further in this INSAG Executive Summary, only the main points are given".

APPENDIX 2

Television Address by Mikhail Gorbachev, 14 May 1986, Moscow

Text released to the 39th World Health Assembly, Geneva, 16 May 1986
Agenda Item 39.1
WHO Reference A39/INF.DOC/10

Good evening, comrades.

As you all know, a misfortune has befallen us—the accident at the Chernobyl nuclear power plant. It has painfully affected Soviet people and caused public anxiety internationally. For the first time ever we have faced the reality of such a sinister force as nuclear energy out of control.

Considering the extraordinary and dangerous nature of what had happened in Chernobyl the Political Bureau took charge of the entire organisation of work to ensure that the breakdown was dealt with as speedily as possible and its consequences limited.

A government commission was formed and immediately left for the scene of the accident, and at the same time a group was set up in the Political Bureau under Nikolai Ivanovich Ryzhkov to solve urgent questions.

All work is being conducted at the present time on a round-the-clock basis. The scientific, technical and economic resources of the entire country have been put to use. Organisations of many Union Ministries and Agencies are operating in the area of the accident under the leadership of ministers, prominent scientists and specialists, along with units of the Soviet army and the Ministry of Internal Affairs.

A huge share of the work and responsibility has been taken upon themselves by the Party, government and economic bodies of the Ukraine and Byelorussia. The operating staff of the Chernobyl nuclear power station are working selflessly and courageously.

So what did happen?

According to specialists, the reactor's output suddenly increased during a scheduled shutdown of the fourth unit. The considerable emission of steam and the subsequent reaction resulted in the formation of hydrogen, which exploded, damaging the reactor and leading to the associated release of radioactivity.

It is as yet early to pass final judgement on the causes of the accident. All aspects of the problem—design, construction, technical and operational—are under the close scrutiny of the government commission.

It goes without saying that when investigation of the causes of the accident is complete, all necessary conclusions will be drawn and measures will be taken to rule out repetition of anything of the sort.

As I have said already, it is the first time that we had encountered such an emergency, one in which it was necessary quickly to curb the dangerous force of the atom that had got out of control and to keep the scale of the accident to the minimum.

The seriousness of the situation was obvious. It was necessary to evaluate it urgently and competently. And as soon as we received reliable initial information it was made available to Soviet people and sent through diplomatic channels to the governments of foreign countries.

On the basis of this information, practical work was launched to deal with the accident and limit its grave aftermath.

In the situation that had arisen we considered it our top priority duty, a duty of special importance, to ensure the safety of the population and provide effective assistance to those who had been affected by the accident.

The inhabitants of the settlement near the power station were evacuated within a matter of hours and then, when it had become clear that there was a potential threat to the health of people in the surrounding zone, they also were moved to safe areas.

All this complicated work required the utmost speed, organisation and precision.

Nevertheless, the measures that were taken failed to protect many people. Two died at the time of the accident—Vladimir Nikolayevich Shashenok, an adjuster of automatic systems, and Valery Ivanovich Khodemchuk, an operator of the nuclear power plant. As of today 299 people are in hospital diagnosed as suffering from radiation sickness to varying degrees. Seven of them have died. Every possible treatment is being given to the rest. The best scientific and medical specialists of the country, and specialised clinics in Moscow and other cities are taking part in treating them and have at their disposal the most modern means of medicine.

On behalf of the Central Committee of the CPSU and the Soviet Government I express profound condolences to the families and relatives of the deceased, to the work collectives, to all who have suffered from this misfortune, and who have experienced personal losses. The Soviet Government will take care of the families of those who died and who were injured.

The inhabitants of the areas that have cordially welcomed the evacuees deserve the highest appreciation. They responded to the misfortune of their neighbours as though it was their own, and in the best traditions of our people displayed consideration, responsiveness and attention.

The Central Committee of the CPSU and the Soviet Government are receiving thousands upon thousands of letters and telegrams from Soviet people and also from people abroad expressing sympathy and support for the victims. Many Soviet families are prepared to take children into their homes for the summer and are offering material help. We have had numerous requests from people to be sent to work in the area of the accident.

These manifestations of humaneness, genuine humanism, and high moral standards cannot but move every one of us.

Assistance to people, I repeat, remains our top priority task.

At the same time we are working energetically at the power station itself and in the surrounding area to limit the scale of the accident. In the most difficult conditions it proved possible to extinguish the fire and prevent it from spreading to the other power units.

The staff of the station shut down the other three reactors and made them safe. They are being constantly monitored.

A stern test has been and is being passed by all concerned, firemen, transport and building workers, medical personnel, special chemical protection units, helicopter crews and other detachments of the Ministry of Defence and the Ministry of Internal Affairs.

In these difficult conditions much depended on a correct scientific evaluation of what was happening, because without such an evaluation it would have been impossible to work out and apply effective measures to cope with the accident and its aftermath. Our prominent scientists from the Academy of Sciences, leading specialists from the Union Ministries and Agencies, the Ukraine and Byelorussia are successfully coping with this task.

I must say that people have acted and are continuing to act heroically, selflessly.

I think we will have an opportunity later to name these courageous people and assess their exploits worthily.

I have every reason to say that, despite the utter gravity of what happened, the damage has turned out to be limited, owing to a decisive degree to the courage and skill of our people, their loyalty to duty, and the concerted way in which everybody taking part in dealing with the aftermath of the accident is acting.

This task, comrades, is being solved not only in the area of the nuclear power station itself but also in scientific institutes, and at many of the country's enterprises which are supplying everything required by those who are directly engaged in the difficult and dangerous struggle to cope with the accident.

Thanks to the effective measures taken, it is possible to say today that the worst is past. The most serious consequences have been averted. Of course, the end is not yet reached. It is not the time to rest. Extensive and lengthy work still lies ahead. The level of radiation in and immediately around the station still remains dangerous to human health. As of today, therefore, the top-priority task is to deal with the effects of the accident. A large-scale programme for the deactivation of the territory of the electric power station and the settlement, of buildings and structures has been drawn up and is being implemented. The necessary manpower, and material and technical resources have been assembled for that purpose. In order to prevent the radioactive contamination of the ground water measures are being taken at the site of the station and in the adjacent area.

Organisations of the meteorological service are constantly monitoring the radiation situation at the surface of the ground, water, and in the atmosphere. They have at their disposal the necessary technical systems and are using specially equipped planes, helicopters and ground monitoring stations.

It is absolutely clear that all these operations will take much time and will require no small efforts. They must be carried out meticulously in a planned and organised manner. The area must be restored to a state that is absolutely safe for the health and normal life of people.

I cannot fail to mention one more aspect of this affair. I mean the reaction abroad to what happened at Chernobyl. In the world as a whole, and this should be emphasised, the misfortune that befell us and our actions in this complicated situation have been treated with understanding.

We are profoundly grateful to our friends in socialist countries who have shown solidarity with the Soviet people at a difficult moment. We are grateful to the political and public figures in other States for sincere sympathy and support.

We are warmly appreciative of foreign scientists and specialists who have shown readiness to come up with assistance in overcoming the consequences of the accident. I would like to note the participation of the American doctors Robert Gale and Paul Terasaki in the treatment of the affected persons and to express gratitude to the business circles of those countries that promptly reacted to our request for the purchase of certain types of equipment, materials and medicines.

We are conscious of the objective attitude to the events of the Chernobyl nuclear power station on the part of the International Atomic Energy Agency (IAEA) and its Director-General, Hans Blix.

In other words, we highly appreciate the sympathy of all those who treated our trouble and our problems with an open heart.

But it is impossible to ignore and make no political assessment of the response to the event at Chernobyl by the governments, political figures and the mass media in certain NATO countries, especially the USA.

They launched an unrestrained anti-Soviet campaign. It is difficult to imagine what was said and written then—"Thousands of Casualties", "Mass Graves of the Dead", "Desolate Kiev", "The Entire Land of the Ukraine Has Been Poisoned", and so on and so forth.

Generally speaking, we faced a veritable mountain of lies—most dishonest and malicious lies. It is unpleasant to recall all this, but it should be done. The international public should know what we had to face. This should be done to find the answer to the question: What, in actual fact, was behind that highly immoral campaign?

Its organisers, to be sure, were not interested in either true information about the accident or the fate of the people at Chernobyl, in the Ukraine, in Byelorussia, in any other place, any other country. They needed a pretext to exploit in the attempt to defame the Soviet Union and its foreign policy, to lessen the impact of Soviet proposals on the termination of nuclear tests and on the elimination of nuclear weapons, and at the same time to dampen the growing criticism of US conduct on the international scene and its militaristic course.

Bluntly speaking, certain Western politicians were after very definite aims—to undermine the possibilities of balancing international relations, to sow new seeds of mistrust and suspicion towards the socialist countries.

All this made itself felt clearly during the meeting of the leaders of "The Seven" held in Tokyo not so long ago. What did they tell the world, what dangers did they warn mankind of? Of Libya groundlessly accused of terrorism, and also of the Soviet Union which, it appears, failed to provide them with "full" information about the accident at Chernobyl. And not a word about the most important thing—how to stop the arms race, how to rid the world of the nuclear threat. Not a word in reply to the Soviet initiatives, to our specific proposals on the termination of nuclear tests, on ridding mankind of nuclear and chemical weapons, on reducing conventional arms.

How should all this be interpreted? One involuntarily gets the impression that the leaders of the capitalist powers who gathered in Tokyo wanted to use Chernobyl as a pretext for distracting the attention of the world public from all those problems that make them uncomfortable, but are so real and important for the whole world.

The accident at the Chernobyl station and the reaction to it have become a kind of a test of political morality. Once again two different approaches, two different lines of conduct, were revealed for everyone to see.

The ruling circles of the USA and their most zealous allies—I would like to mention especially the Federal Republic of Germany—regarded the mishap only as another opportunity to put up additional obstacles holding back the development and deepening of the current East–West dialogue, slow though its progress is, and to justify the nuclear arms race.

What is more, an attempt has been made to prove to the world that talks and, even more, agreements with the USSR are impossible, and thus to give the green light to further military preparations.

Our attitude to this tragedy is absolutely different. We realised that it is another tolling of the bell, another grim warning that the nuclear era necessitates new political thinking and a new policy.

This has strengthened still more our conviction that the lines of foreign policy worked out by the 27th Congress of the CPSU are correct and that our proposals for the complete elimination of nuclear weapons, the ending of nuclear explosions, and the creation of an all-embracing system of international security meet those inexorably stringent demands which the nuclear age makes on the political leadership of all countries.

As to the "lack" of information, around which a special campaign has been launched, and a campaign of a political content and nature at that, this charge in the given case is a spurious one. The following facts confirm that this, indeed, is so. Everybody remembers that it took the US authorities ten days to inform their own congress and months to inform the world community about the tragedy that took place at Three Mile Island atomic power station in 1979.

I have already said how we acted.

All this enables one to judge who best approaches the matter of informing their own people and foreign countries.

But the essence of the matter is different. We hold that the accident at Chernobyl, like the accidents at United States, British and other atomic power stations, poses to all States very serious problems which require a responsible attitude.

Over 370 atomic reactors now function in different countries. This is reality. The future of the world economy can hardly be imagined without the development of atomic power. Altogether 40 reactors with an aggregate capacity of over 28 million kilowatts now operate in our country. As is known, mankind derives a considerable benefit from atoms for peace.

But it stands to reason that we are all obliged to act with still greater caution, to concentrate the efforts of science and technology to ensure the safe harnessing of the great and formidable powers contained in the atomic nucleus.

The indisputable lesson of Chernobyl to us is that, in the further development of the scientific and technical revolution, questions of the reliability and safety of equipment, questions of discipline, order and organisation assume priority importance. The most stringent demands are needed everywhere and in everything.

Furthermore, we consider it necessary to call for a serious deepening of co-operation in the framework of the International Atomic Energy Agency (IAEA). What steps could be considered in this connection?

First, creating an international regime for the safe development of nuclear power on the basis of close co-operation between all nations dealing with nuclear power engineering. A system of prompt warning and supply of information in the event of accidents and faults at nuclear power stations, specifically when this is accompanied by the escape of radioactivity, should be established in the framework of this regime. Likewise it is necessary to adjust the international machinery, both on a bilateral and a multilateral basis, to ensure the speediest rendering of mutual assistance when dangerous situations emerge.

Second, for discussion of the entire range of matters, it would be justifiable to convene a highly authoritative specialised international conference in Vienna under IAEA auspices.

Third, in view of the fact that IAEA was founded as long ago as 1957 and that its resources and staff are not in keeping with the level of development of present-day nuclear power engineering, it would be expedient to enhance the role and possibilities of that unique international organisation. The Soviet Union is ready for this.

Fourth, it is our conviction that the United Nations Organisation and its specialised institutions, such as the World Health Organisation (WHO) and the United Nations Environment Programme (UNEP), should be involved more actively in the effort to ensure safe development of peaceful nuclear activity.

For all that, it should not be forgotten that in our world, where everything is interrelated, there exist, alongside problems of atoms for peace, also problems of atoms for war. This is the main thing now. The accident at Chernobyl showed again what an abyss will open if nuclear war befalls mankind. For inherent in the nuclear arsenals stockpiled are thousands upon thousands of disasters far more horrible than the Chernobyl one.

Given the increased attention to nuclear matters, the Soviet Government, having considered all the circumstances connected with the safety of its people and the whole of humanity, has decided to extend its unilateral moratorium on nuclear tests until August 6 of this year, that is until the date on which, more than 40 years ago, the first atomic bomb was dropped on the Japanese city of Hiroshima, as a result of which hundreds of thousands of people perished.

We urge the United States again to consider with the utmost responsibility the measure of danger looming over mankind, and to heed the opinion of the world community. Let those who are at the head of the United States show by deeds their concern for the life and health of people.

I confirm my proposal to President Reagan to meet without delay in the capital of any European State that will be prepared to accept us or, say, in Hiroshima and to agree on a ban on nuclear testing.

The nuclear age forcefully demands a new approach to international relations, and the pooling of efforts of States with different social systems, for the sake of putting an end to the disastrous arms race and of effecting a radical improvement in the world political climate. Broad horizons will then be cleared for fruitful co-operation of all countries and peoples, and all men on earth will gain from that.

APPENDIX 3

Cancer Incidence in Eastern Europe and Scandinavia

FROM the map in Fig. 35 showing the initial direction of the radioactive plume from Chernobyl to Forsmark in Sweden, it is clear that the major fall-out levels occurred in Eastern Europe and Scandinavia. The public concern about iodine-131 and caesium-137 has already been mentioned in Chapter 7, and published estimates of excess cancer deaths have been quoted in Chapter 9. However, before an assessment can be made of future excess deaths from cancer of the thyroid and from leukaemia—which are the malignancies most often mentioned in relation to radiation-induced cancer—the current incidence patterns must be known. This appendix gives a summary of the latest available data in international publications for the Soviet Union. It also compares this with data for other Eastern European countries and for Scandinavia.

The most recent readily available data for the Soviet Union is for the years 1969–1971, and is for all fifteen Soviet Socialist Republics, including the Ukraine and Byelorussia. It was published in 1983 by the International Agency for Research on Cancer as a supplement to the 1976 Volume III of the IARC "Cancer Incidence in Five Continents", which contained no data from the Soviet Union. Volume IV was published in 1982, but it also contained no Soviet data.

IARC Volumes III and IV contain cancer incidence data for individual International Classification of Disease Numbers, some 100 in all. However, the Soviet data is only for the following groups of ICD Numbers:

140	Lip
141–149	Mouth and throat (termed other oral cavity and pharynx)
150	Oesophagus
151	Stomach
154	Rectum
161	Larynx
162	Lung (termed bronchus and trachea)

172–173 Skin
200–209 Lymphatic and haematopoietic cancer
and Other and unspecified cancer

This means that cancers of the thyroid (ICD Number 194) are grouped with Other and unspecified cancer, and that leukaemia (ICD Number 204, lymphatic; 205, myeloid; 205, monocytic; 207, other leukaemia) is grouped with Lymphatic and haematopoietic cancer (ICD Number 200, lymphosarcoma etc.; 201, Hodgkin's disease; 202, other reticuloses; 203, multiple myeloma; 208, polycythemia vera; 209, myelofibrosis). However, the consistently low incidence of cancer of the thyroid (usually in the ranges 60–200 per million male population and 100–250 per million female population) for other Eastern European countries is probably a good indication of that which occurs in the Soviet Union. It is not so easy to estimate the incidence of leukemia in the Soviet Union though, because the incidence of ICD numbers 200–209 cancers appears generally lower in the Soviet Union than in the rest of Eastern Europe and it is not clear if this is due to either leukaemia or to lymphoma, or to both.

Any cancer incidence comparisons must be those of the same statistical parameters and for the Soviet Union to be included this must be crude average annual incidence rates. Age-specific incidence rates are for different age groups in the IARC publication for the Soviet Union and Volumes III and IV for the rest of the world. For the Soviet Union the intervals are every 10 years in age, whereas for Volumes III and IV they are for every 5 years.

The first table compares the average annual incidence (i.e. the number of new cancer cases) for Eastern Europe and for Scandinavia, for males, for cancer of the thyroid, for all leukaemias (ICD Numbers 204–207), and for cancers of the lung and stomach, which are high-incidence cancers in these areas. Lung cancer incidence exceeds that of stomach cancer in all areas quoted, except Romania and in the Soviet Union. In one part of Hungary, the incidence of these two cancers in males is approximately equal. It is also interesting to note from this table that for Poland, lung cancer incidence is by far the highest in the only two city areas for which data is available, Cracow and Warsaw.

From the tables it is seen that:

—The incidence of cancer of the thyroid, although very low, is always greater for women than for men.

—The incidence of leukaemia is usually double that of cancer of the thyroid for women, and for men may be six times greater than for cancer of the thyroid.

Other interesting features are:

—In men, stomach cancer in the Soviet Union and in Romania has a noticeably higher incidence than lung cancer.

—In women, the incidence of cancer of the cervix in the Soviet Union and in Romania is slightly higher than the incidence of cancer of the breast, whereas in many other countries the incidence of breast cancer is much higher than that of the cervix.

—For those areas considered in the tables, the incidence in women of cancer of the stomach is by far the highest in Byelorussia, 403 per million female population. For

Cancer Incidence in Males for Selected Populations

Geographical area and registration years	Number of cancers per million male population			
	Thyroid	Leukaemia	Stomach	Lung
Scandinavia				
Sweden, 1971–75	21	125	316	395
Denmark, 1973–76	12	111	266	757
Norway, 1973–77	19	98	321	374
Finland, 1971–76	18	81	328	853
Eastern Europe				
German D.R., 1973–77	11	86	455	835
(a) Czechoslovakia, 1973–77	8	62	336	526
(b) Hungary, 1973–77	6	33	356	327
(c) Hungary, 1973–77	9	58	552	719
(d) Poland, 1973–77	12	97	417	474
(e) Poland, 1973–77	12	65	328	698
(f) Poland, 1973–74	6	40	353	470
(g) Poland, 1973–77	19	65	423	481
(h) Poland, 1973–77	9	78	355	780
(i) Poland, 1973–77	1	70	359	473
(j) Romania, 1974–78	4	63	504	414
(k) Yugoslavia, 1973–76	17	69	465	613
Soviet Union				
Ukraine, 1969–71	not known		421	402
Byelorussia, 1969–71	not known		535	247
All USSR, 1969–71	not known		478	360

Cancer Incidence in Females for Selected Populations

Geographical area and registration years	Number of cancers per million female population				
	Thyroid	Leukaemia	Stomach	Breast	Cervix
Scandinavia					
Sweden, 1971–75	54	90	202	883	168
Denmark, 1973–76	22	83	176	853	284
Norway, 1973–77	55	76	206	731	224
Finland, 1971–76	50	65	244	535	112
Eastern Europe					
German, D.R., 1973–77	25	71	318	597	384
(a) Czechoslovakia, 1973–77	23	42	192	378	165
(b) Hungary, 1973–77	20	23	179	258	134
(c) Hungary, 1973–77	20	42	325	292	187
(d) Poland, 1973–77	22	47	215	240	210
(e) Poland, 1973–77	25	65	238	352	246
(f) Poland, 1973–74	14	35	196	288	250
(g) Poland, 1973–77	21	57	286	257	163
(h) Poland, 1973–77	34	68	236	532	270
(i) Poland, 1973–77	9	46	155	205	179
(j) Romania, 1974–78	19	42	303	379	355
(k) Yugoslavia, 1973–76	28	52	275	444	213
Soviet Union					
Ukraine, 1969–71	not known		291	224	278
Byelorussia, 1969–71	not known		403	158	175
All USSR, 1969–71	not known		382	187	260

Notation

(a) Western Slovakia; (b) Szabolcs-Szatmar County; (c) Vas County; (d) Cieszyn Area; (e) Cracow City; (f) Katowice District; (g) Nowy Sacz; (h) Warsaw City; (i) Warsaw Rural Areas; (j) Cluj County; (k) Slovenia.

men in Byelorussia, it is 535 per million male population, which is comparable to the figure of 504 in Romania.

—The four pie charts which follow show the pattern of cancer incidence for males and females, in the Ukraine and in Byelorussia, using all available data from the IARC publication. The two bar charts for males and females compare the Soviet data for lymphatic and haematopoietic cancer (ICD Numbers 200–209). Any study of excess cancer incidence (as distinct from deaths, since all cancer patients do not die from their malignant disease) due to the radioactive releases from the Chernobyl accident will have to take such cancer incidence patterns into consideration, if a statistically significant excess is to be demonstrated.

BYELORUSSIA, 1969-1971, MALES

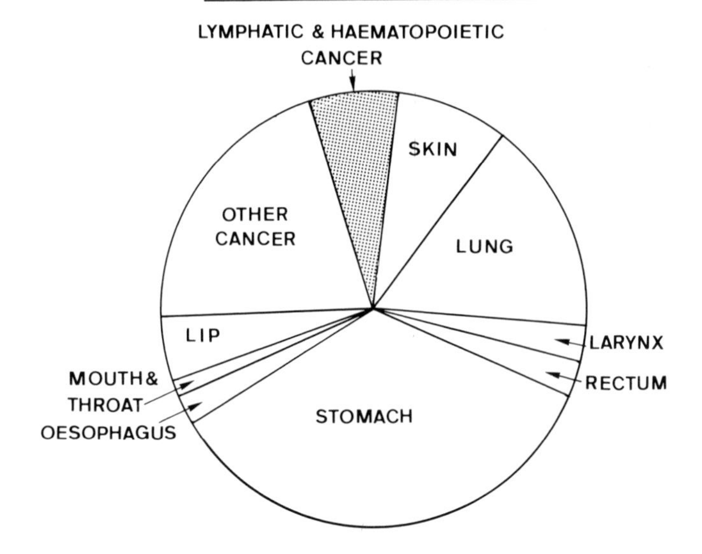

TOTAL NUMBER OF MALES = 19, 206

BYELORUSSIA, 1969-1971, FEMALES

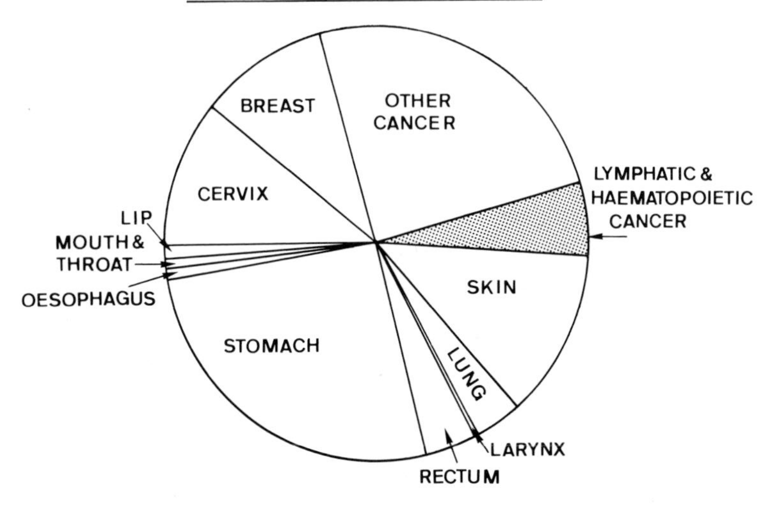

TOTAL NUMBER OF FEMALES = 22,739

UKRAINE, 1969–1971, MALES

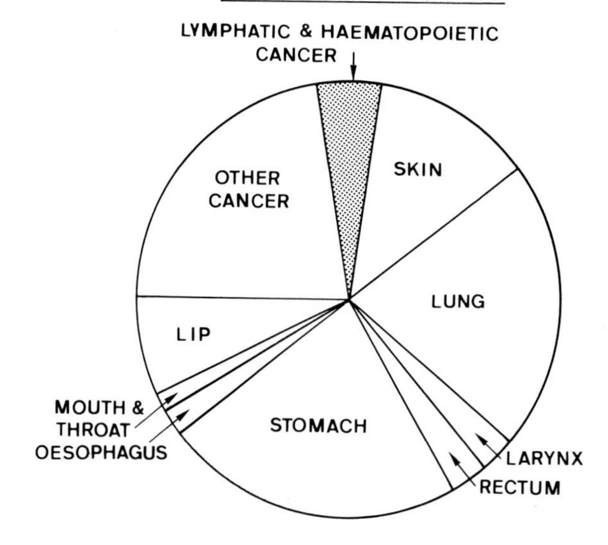

TOTAL NUMBER OF MALES = 118,562

UKRAINE, 1969–1971, FEMALES

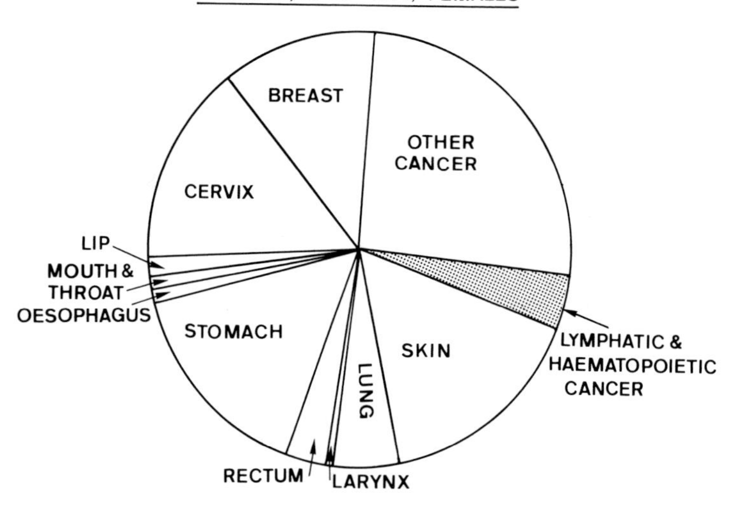

TOTAL NUMBER OF FEMALES = 144,904

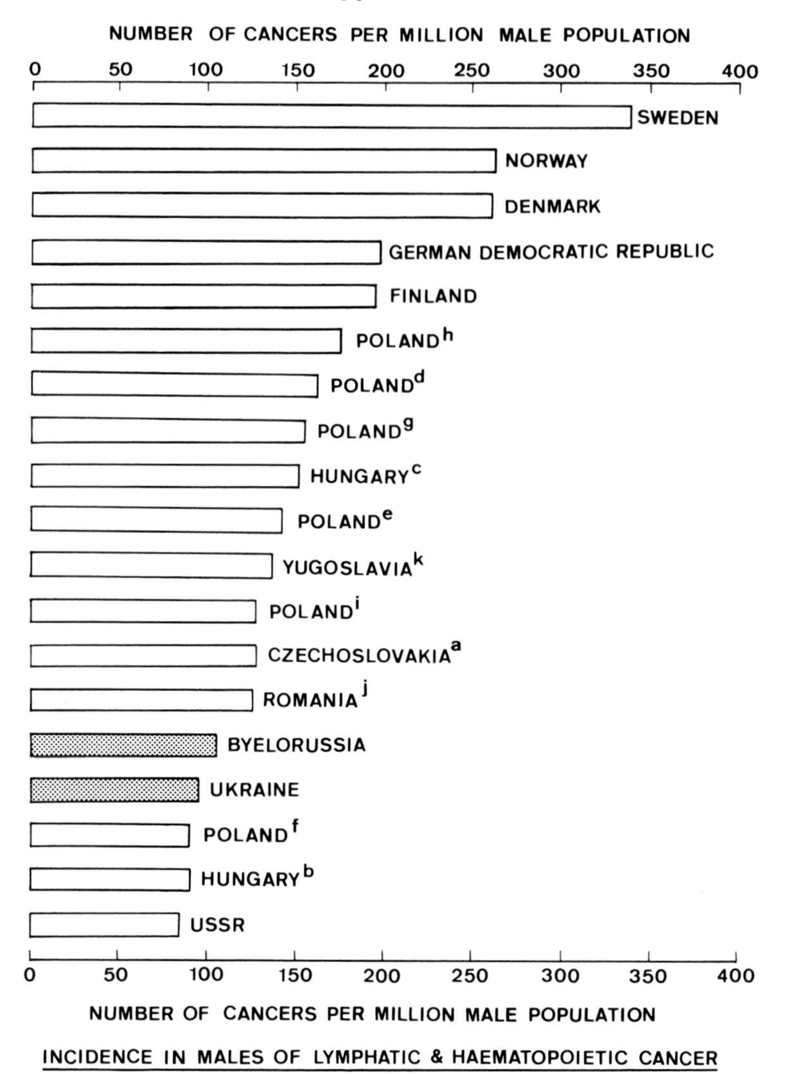

See p. 205 for notation

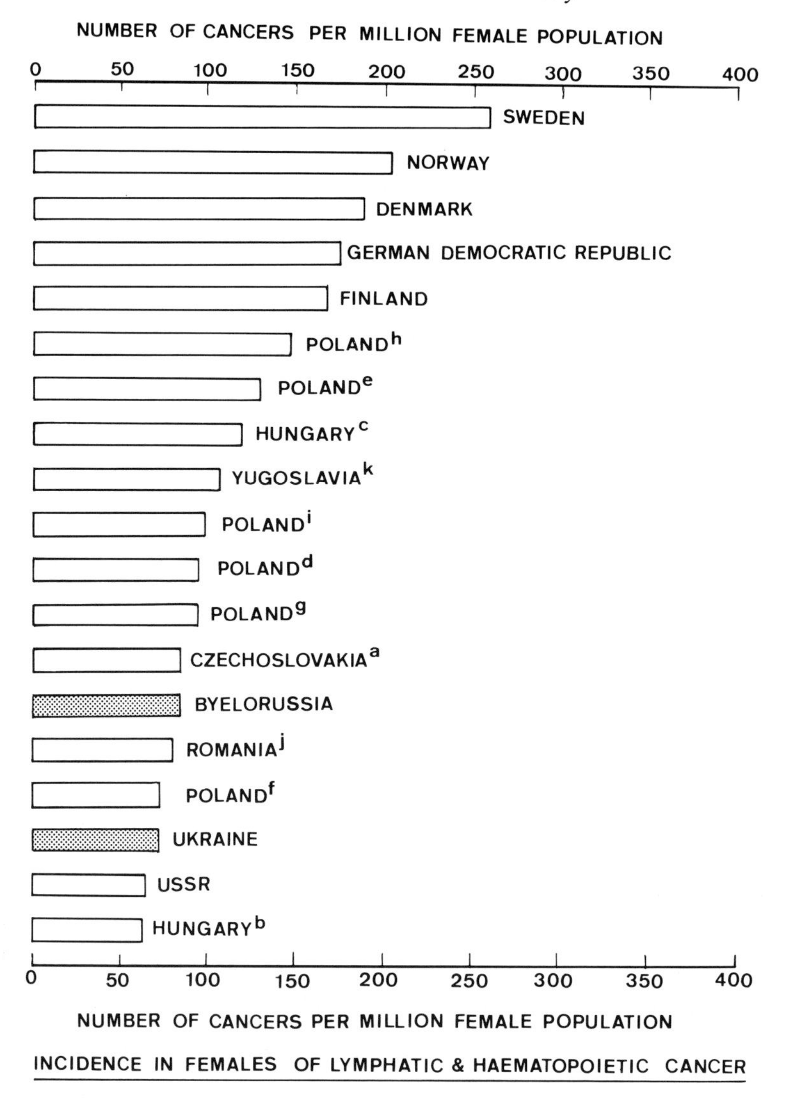

INCIDENCE IN FEMALES OF LYMPHATIC & HAEMATOPOIETIC CANCER

See p. 205 for notation

APPENDIX 4

Soviet Manual for the Medical Treatment of Radiation Victims

THE manual was presented to the author by Professor Angelina Guskova on 12 April 1987 during her visit to London for the Institute of Biology seminar on Chernobyl. This is the standard reference work in the Soviet Union on the treatment of radiation injuries.

Manual for the Organisation of Medical Treatment of Persons who have been affected by Ionising Radiation, 1986, L. A. Ilyin (Editor), Energoatomizdat (scientific edition), Moscow (publishing approval given on 5 August 1986, print run of 4650 copies, Russian language edition)

РУКОВОДСТВО

по организации медицинского обслуживания лиц, подвергшихся действию ионизирующего излучения

The contents of the manual are:

Chapter 1 *The main biological effects of radiation and the general principles of their studies* by A. K. Guskova, V. I. Kiruyshkin and M. M. Kosenko.

Chapter 2* *Methodology of special investigations of persons who work with sources of ionising radiation.*

Chapter 3* *Principles of investigations for the prevention and treatment of disease in persons who work in the production of different kinds of sources of radiation,*

3.1 *Uranium,*

3.2 *Polonium,*

3.3 *Plutonium.*

Chapter 4 *Organisation of monitoring the limited part of the population who are working in an environment of a higher than ordinary level of radiation* by V. I. Kiruyshkin, A. K. Guskova and M. M. Kosenko.

---

* Chapters 2 and 3 are each written by some 15 authors. I would also like to thank Mr I. Peskov of *TASS* for help with translation of this list of contents.

APPENDIX 5

The Chernobyl Victims' Fund

A DISASTER fund for the victims of the Chernobyl accident was set up in the Soviet Union after 26 April 1986 and by 1 September the Soviet people had contributed 480 million roubles to the Chernobyl relief fund in account 904 of the State Bank. Additionally, by 1 September, a similar relief fund in the Foreign Trade Bank held 1.3 million roubles in foreign currency. Bank accounts for the fund are also available in several countries, for example in the United Kingdom the account is 141 505 CRF at the Moscow Narodni Bank in London. Donations to the fund have come from many and varied sources, but one of the most well publicised was a 30 May charity concert in Moscow's Olympic Sports Centre. This concert was called Chernobyl Aid and raised some 90,000 roubles. Popular Soviet singers such as Alla Pugachoyva (*far right in the photograph*) and Alexander Gradsky took part, together with the groups Recital, Autograph and Cruise.

It has also been reported by Novosti that the total cost of the evacuation, decontamination, entombment of the reactor and other necessary procedures will cost some two billion roubles. The exchange rate for £-sterling is, at the time of publication, some £1.02=1 rouble.

Soviet singer Alla Pugachyova (*right*) and other pop artistes who held a charity concert at the Olympiysky sports complex in Moscow, the proceeds of which has gone to the victims of the Chernobyl nuclear accident.

Courtesy of TASS

*Glossary**

Absorbed dose. Exposure is a quantity relating to ionisation in *air* and is used to describe a property of X-rays or gamma rays emitted from an external therapy machine. It describes the radiation output in air from the machine or, when expressed in rontgens per minute (R/min), the output rate in air. It does not, however, describe the energy imparted to an irradiated material, and therefore cannot be used to specify the radiation energy absorbed by a target volume within a patient. The quantity required is the absorbed dose (D), and its special unit is the *rad.*

$$D = \frac{\Delta E_d}{\Delta m}$$

where ΔE_d is the energy imparted by ionising radiation to the matter in a volume element and Δm is the mass of matter in that volume element (1 rad=100 erg/g).

Absorbed dose and exposure may be related using the formula

$$\text{Absorbed dose} = \text{Exposure} \times f$$

where f is a factor dependent upon the quality of the radiation beam and the material being irradiated.

The SI unit of absorbed dose is the gray (Gy), which is equal to 1 joule/kg and thus 1 Gy = 100 rad.

For the absorbed dose, there are the following relationships (c = centi, m = milli):

1 rad = 0.01 Gy or 1 cGy
1 mrad = 0.01 mGy
10 mrad = 0.1 mGy
100 mrad = 1 mGy
500 rad = 5 Gy or 500 cGy

Activity. The curie (Ci) was first suggested as a unit of radioactivity in 1910, but was only defined for use with radon: "The quantity of radon in radioactive equilibrium with 1 gram of radium." It was later extended to include all radioactive isotopes as "a unit of activity which gives 3.700×10^{10} disintegrations per second". The new SI unit of activity is the becquerel (Bq), which equals one disintegration per second.

* See Additional Glossary, p. 239–243.

For activity, there are the following relationships (m = milli, M = mega, G = giga, T = tera):

$$1\ \text{Ci} = 37 \times 10^{9}\ \text{Bq}$$
$$1\ \text{Bq} = 27.03 \times 10^{-12}\ \text{Ci}$$
$$2\ \text{mCi} = 74\ \text{MBq}$$
$$10\ \text{mCi} = 370\ \text{MBq}$$
$$100\ \text{mCi} = 3.7\ \text{GBq}$$
$$1000\ \text{Ci} = 37\ \text{TBq}$$

Acute radiation syndrome. When the whole body or a major part of it is exposed to a large acute dose of penetrating radiation (gamma or X-ray, or neutrons) a pattern of disease develops known as the *acute radiation syndrome*. The underlying cause of the pattern is the radiosensitivity of three organs, all of which play an essential part in sustaining life. According to the dose delivered, the haematopoietic tissues, the lining of the small intestine and finally the central nervous system are affected. The more of the trunk included in the exposure the worse will be the illness, because of the location there of both the small intestine and a large portion of the haematopoietic tissues.

Age-specific cancer incidence rate. An age-specific incidence rate refers to a population in a specified age range, usually a 5-year or 10-year range. Thus for the age range 40–45 years, the annual age-specific incidence rate per 100,000 population for males is equal to:

$$\frac{\text{Number of new cancer cases in males aged 40–45 years registered in year Y}}{\text{Average number of males aged 40–45 at risk in year Y}} \times 100{,}000$$

A similar rate can be defined for females aged 40–45 years. The advantage of the age-specific incidence rate over the crude incidence rate is that any peculiarities in the cancer incidence pattern which are related to particular age groups can be shown. Variations with age would not be apparent using only a crude incidence rate.

Allogenic. Of a different genetic constitution. In bone marrow transplants, this applies to grafts between one individual and another of the same species. The new bone marrow is taken from a donor.

Alpha particle. A particle consists of two protons and two neutrons. It is emitted during the decay of some radioactive isotopes. A helium nucleus also consists of two protons and two neutrons.

Atom. The smallest portion of an element that can combine chemically with other atoms.

Atomic number. The number of protons in a nucleus of an atom.

Autologous. Of the same genetic constitution.

Becquerel unit. *See* **Activity** and **Radiation units: SI and non-SI: conversion factors.**

Beta particle. An electron emitted during the decay of some radioactive isotopes. If the electron has a positive electric charge, then it is called a positron.

Boron. A powerful absorber of neutrons, which is used in the construction of nuclear reactor control rods.

Caesium. The radioactive isotope caesium-137 is a fission product, which emits beta particles when it decays with a 30.2 year half-life. It is used in cancer treatment (radiotherapy) in the form of caesium needles which can be implanted into a tumour, or in caesium sources placed within the vagina and uterus to treat cancer of the cervix and cancer of the uterus.

Cancer incidence. *See* **Crude cancer incidence rate** and **Age-specific cancer incidence rate.**

CEGB. Central Electricity Generating Board in the United Kingdom.

Centi. Prefix for one-hundredth, i.e. 10^{-2}.

Chromosome. One of the several small dark-staining and more or less rod-shaped bodies which appear in the nucleus of a cell at the time of cell division. They contain the genes, or hereditary factors, and are constant in number in each species. The normal number in man is 46.

Collective effective dose equivalent. This is sometimes termed *collective dose* or *collective dose equivalent.* The quantity is obtained by multiplying the average effective dose equivalent by the number of persons exposed to a given source of radiation. It is expressed in man-sievert units. Its use is illustrated from a commentary in *The Lancet* of 13 September 1986, following the 25–29 August meeting. "It was calculated that the 135,000 evacuees had received a total of 16,000 man-sieverts (1.6 million man-rem) in collective dose from external radiation alone, with some 25,000 of those living 3–15 kilometres from the plant receiving average doses of 350–550 millisievert (35–55 rem)." To set these values in context, the average annual radiation dose to the population in the United Kingdom from all sources is less than 2 millisievert and the annual dose limit for a radiation worker is 50 millisievert.

Crude cancer incidence rate. An annual incidence rate is a measure of the new cases of a disease in a particular year. It is usually quoted as a proportion per 100,000 of a defined population at risk, but can also be stated per million or per thousand population at risk. The adjective *crude* refers to the fact that the rate is not modified to take into account such factors as age or reference year. The crude annual incidence rate per 100,000 population for a specified cancer in males is equal to:

$$\frac{\text{Number of new cancer cases in males registered in year Y}}{\text{Average number of males at risk in year Y}} \times 100{,}000$$

A similar rate can be defined for females. A crude incidence may be calculated for individual cancers or for cancer at all sites.

Curie unit. *See* **Activity and Radiation units: SI and non-SI: conversion factors.**

Cytoplasm. The protoplasm of a cell surrounding the nucleus.

Derived action levels. These are preplanned radiation levels which are used in a radiation accident situation to decide what action is to be taken. There are at present no internationally agreed derived action levels for food such as milk and leafy vegetables, or for evacuation populations. At the 25–29 August 1986 meeting, Professor Ilyin stated that in the Soviet Union in 1969 criteria were drawn up for evacuation of the population in the event of radioactive releases. These were:

Radiation level (Dose equivalent)	Action
Up to 25 rem	No evacuation
25–75 rem	Evacuation obligatory but account to be made of the existing conditions
Over 75 rem	Basis for emergency evacuation

Deuterium. Heavy hydrogen, which has a nucleus consisting of one proton plus one neutron. It is sometimes termed hydrogen-2. Hydrogen-1 is ordinary hydrogen and has only a single proton in its nucleus.

Discovery of radium. It was Becquerel's discovery of radioactivity that led Pierre (1859–1906) and Marie Curie (1867–1934) to announce the discovery of radium on 26 December 1898. The Curies had a more suitable radiation measuring device than a photographic plate – an electrometer equipped with an ionisation chamber – and they demonstrated that the intensity of the radiation was proportional to the amount of uranium. Many substances were studied and it was eventually found that the uranium mineral pitchblende showed a higher radiation intensity than could be explained simply by the presence of radioactive uranium and thorium compounds. This was followed by the discovery of the radioactive element polonium, associated with the bismuth extract of the ore, and finally radium, associated with the barium extract. However, although a knowledge of the new element, radium, had been gained, it was an immense problem to refine it from pitchblende ore in any quantity and in 1902 it was still considered an achievement to obtain a yield of 260 mg of radium bromide per ton of ore. However, such processing problems are now only an historical memory.

Discovery of X-rays. X-rays were discovered in Würzburg, Germany, on 8 November 1895 by Wilhelm Röntgen (1845–1923) when he was investigating the phenomenon of the passage of an electrical discharge from an induction coil through a partially evacuated glass tube. Although the tube was covered in black paper and the whole room was in complete darkness, he observed that a paper screen covered with a fluorescent material became illuminated. These strange new rays were also found to penetrate other objects, such as a wooden plank, a thick book and metal sheets. X-rays had been discovered and the first public announcement was made that year in December under the title "On a new kind of ray". Generally, this discovery was received with acclaim and headlines appeared entitled "Illuminated tissue", "Electrical photography through solid bodies" and other such unusual descriptions, though at least one magazine remarked on the "revolting indecency of seeing other people's bones". X-ray pictures of hands, purses, Egyptian mummies and small animals were soon taken for demonstration purposes in many countries.

Dose. This is a general term for a quantity of radiation, but since there are several types of defined dose, e.g. *absorbed dose*, *collective effective dose equivalent*, the term *dose* should be used more precisely.

Dose-effect relationships. The following table, *Dose–Effect Relationships for Acute Ionising Radiation Exposure*, is taken from a paper by J. Geiger in the *Journal of* *the American Medical Association* (1 August 1986, p. 610). The author comments that recently questions have arisen about both the upper and lower ranges, and that a re-analysis of the data from Hiroshima and Nagasaki is being undertaken.

Dose rem (sievert)	Clinical illness	Percentage surviving with treatment
15–50 (0.15–0.5)	Asymptotic, may be chromosomal aberrations	100%
100–200 (1–2)	Nausea and vomiting, bone marrow suppression	100%
300–400 (3–4)	Severe leukopenia, thrombocytopenia and epilation	50%
600–1000 (6–10)	Gastrointestinal syndrome	0–10%
1000–5000 (10–50)	Acute encephalopathy and cardiovascular collapse	0%

Dose equivalent. In an attempt to measure the maximum amounts of ionising radiations which persons could safely receive, a quantity called the *relative biological effectiveness* (RBE) was proposed in which selected values of RBE were multiplied by the energy of the ionising radiation per unit mass. The unit of RBE dose was the *rem*, which meant rontgen equivalent man. With the introduction of the *rad* as the unit of absorbed dose, a quantity called *dose equivalent, H*, replaced the previous RBE dose in 1962, where H is weighted absorbed dose given by

$$H = DQN.$$

Here D is the absorbed dose, Q is one weighting factor which is a quality factor for the ionising radiation, and N is the product of all other weighting factors that might modify the potentially harmful biological effects of the absorbed dose D. At present the value of N is taken as 1, but future information may cause this to be modified. The quality factor, Q, has the following recommended values:

$Q = 1$ for X-rays, gamma rays and electrons
$Q = 10$ for neutrons and protons
$Q = 20$ for alpha particles

Thus the quality factor is extremely important when the effects of different radiations have to be combined; for example, X-rays and neutrons. In such an instance, the biological response *must* be taken into account.

The SI unit of dose equivalent is the sievert (Sv), and it has the following relationships to the rem (μ = micro, m = milli)

$$1 \text{ rem} = 10^{-2} \text{ Sv}$$
$$1 \text{ Sv} = 100 \text{ rem}$$
$$1 \text{ mrem} = 10 \ \mu\text{Sv or } 0.01 \text{ mSv}$$
$$10 \text{ mrem} = 0.1 \text{ mSV}$$

$$100 \text{ mrem} = 1 \text{ mSv}$$
$$1 \text{ rem} = 10 \text{ mSv}$$

Electron. Elementary particle with a negative charge equal and opposite to that of a proton. Its mass is only 1/1836th of a proton.
Encephalopathy. Any disorder of the brain.
Enriched uranium. Uranium in which the content of the radioactive isotope uranium-235 has been increased above its natural value of 0.7% by weight. Uranium fuel enrichment for the RBMK1000 reactors is 2%, but following the accident this is to be increased to 2.4%. This will significantly reduce the void coefficient of the reactor and provide added safety.
eV. Electron volt. This equals the energy gained by an electron in passing through a potential difference of 1 volt.
Exposure. The rontgen unit of 1937 was defined as: "That amount of X- or gamma radiation such that the associated corpuscular emission per 0.001293 gram of air produced in air ions carrying 1 electrostatic unit of charge of either sign." 0.001293 g is the weight of 1 cm^3 of air at normal temperature and pressure. This definition was later revised and the *rontgen* was termed the special unit of a quantity called exposure (X) where

$$X = \frac{\Delta Q}{\Delta M}$$

ΔQ is the sum of all the electrical charges of all the ions of one sign produced in air when all the electrons (negatrons and positrons) liberated by photons in a volume element of air whose mass is Δm are completely stopped in air. The rontgen unit (R) $= 2.58 \times 10^{-4}$ coulomb/kg of air. There is no special name for the SI unit of exposure, but its magnitude is 1 coulomb/kg, which is equivalent to approximately 3.876×10^3 R.

For activity, there are the following relationships (m = milli, M = mega, T = tera):

$$1 \text{ Ci} = 37 \times 10^9 \text{ Bq}$$
$$1 \text{ Bq} = 27.03 \times 10^{-12} \text{ Ci}$$
$$2 \text{ mCi} = 74 \text{ MBq}$$
$$10 \text{ mCi} = 370 \text{ MBq}$$
$$1000 \text{ Ci} = 37 \text{ TBq}$$

Fast neutrons. Conventionally, neutrons with energies in excess of 0.1 MeV, which corresponds to a velocity of about 4,000,000 metres per second. Thermal neutrons at ordinary temperatures have energies of about 0.025 eV, which correspond to a velocity of about 2200 metres per second.
FDA. Food and Drug Administration in the United States of America.
Fermi reactor fuel melting. The following is taken from a 1 July 1986 Congressional Issue Brief, USA. "The Enrico Fermi Atomic Power Plant at Lagoona Beach, Michigan, featured a small sodium-cooled, fast-breeder reactor with a generating capacity of 61 MWe. On 5 October 1966, the fuel in two fuel assemblies melted because of a coolant blockage. There were no injuries and there was no release of radioactivity. The reactor resumed operation in October 1979 after repairs.

However, the accident attracted substantial attention because of fears that the reactor vessel might be breached."

Fission. Nuclear fission is a process in which a nucleus splits into two or more nuclei and energy is released. Fission frequently refers to the splitting of a nucleus of uranium-235 with emission of other neutrons.

Gamma ray. A discrete quantity of electromagnetic energy (sometimes called a *photon* of energy), without any mass or any electric charge. Gamma rays are emitted during the decay of a radioactive isotope and accompany the emission of either an alpha particle or a beta particle.

Gene. *See* **Chromosome.**

Giga. Prefix for hundred thousand millions, i.e. 10^9.

Granulocytes. A type of leucocyte which has a granular cytoplasm.

Gray unit. *See* **Absorbed dose** and **Radiation units: SI and non-SI: conversion factors.**

Haematopoietic. Related to the formation of blood cells. Haematopoietic is spelt alternatively as hemopoietic.

Half-life. The time taken for the activity of a radioactive isotope to lose half its value by decay. Half-lives can vary from fractions of a second to thousands of years.

Heavy water. D_2O, where D is deuterium, i.e. hydrogen-2.

Hyperaemia. An excess of blood in a part of the body.

IAEA. International Atomic Energy Agency.

ICRP. International Commission on Radiological Protection.

INSAG. The International Nuclear Safety Advisory Group. This is an advisory group to the Director General of the IAEA. Its main functions are:

(1) To provide a forum for the exchange of information on generic nuclear safety issues of international significance.
(2) To identify important current nuclear safety issues and to draw conclusions on the basis of results of nuclear safety activities within the IAEA and of other information.
(3) To give advice on nuclear safety issues in which an exchange of information and/or additional efforts may be required.
(4) To formulate, where possible, commonly shared safety concepts.

In December 1986 the IAEA published the INSAG summary report on the Post-Accident Review meeting (25–29 August 1986) on the Chernobyl accident.

Iodine. Iodine accumulates in the thyroid gland and radioactive iodine-131, a fission product, emits beta particles when it decays with a 8.1 day half-life. In medicine, it is used in small activities to diagnose thyroid disorders, and in larger activities to treat cancer of the thyroid.

Ion. *See* **Ionisation.**

Ionisation. A process in which a stable, electrically neutral atom loses an electron. This electron is then termed a negative ion and the now net positivity charged atom is termed a positive ion.

Ionising radiation. Radiation that produces ionisation in atoms. Examples are alpha particles, beta particles, gamma rays, X-rays and neutrons.

Isotopes. A species of atom with the same number of protons (i.e. same *atomic number*) but different numbers of neutrons (i.e. different *mass numbers*). Thus uranium-238 and uranium-236 both have the same atomic number (92), different mass numbers (238 and 236) and different numbers of neutrons (146 and 144). A stable isotope is one which is not radioactive, and thus does not decay.

Kilo. Prefix for thousands, i.e. 10^3.

Kyshtym. The following is taken from a 1 July 1986 Congressional Issue Brief, USA. "Reports of a disaster in the Urals in 1957 or 1958. There have been reports of a major accident in 1957 or 1958 which was claimed to have cost many lives and contaminated a wide area with dangerous levels of radioactivity. The accident presumably occurred in the region of Kyshtym, east of the Ural mountains, the site of the first Soviet plutonium production facility that started in 1947."

Leucocytes. Blood is composed of a fluid called plasma in which are suspended red corpuscles, white cells and platelets. Leucocyte is a term for a white blood cell. In normal blood there are approximately 8000 per cubic millimetre.

Leukaemia. An often fatal disease of the blood-forming organs, characterised by a marked increase in the number of leucocytes and their precursors in the blood. It is sometimes colloquially termed cancer of the blood.

Leukopenia. Deficiency of the number of leucocytes in the blood.

Light water. H_2O, ordinary water, where H is hydrogen-1. *See also* **Deuterium.**

Lymphocytes. A type of leucocyte which has a clear cytoplasm.

Lymphoid. Resembling lymph of lymphatic tissue.

Lymphopenia or Lymphocytopenia. Absolute or relative diminution in the number of lymphocytes present in the blood.

Mass number. The number of protons plus neutrons in a nucleus of an atom.

Mega. Prefix for millions, i.e. 10^6.

Megawatt (MW). One million watts or one thousand kilowatts.

Meltdown. All or part of the solid nuclear fuel in a reactor reaching a temperature in which the fuel cladding and the fuel liquefies and collapses. This is a consequence of overheating of the reactor core.

MeV. Million electron volts.

Micro. Prefix for one-millionth, i.e. 10^{-6}.

Milli. Prefix for one-thousandth, i.e. 10^{-3}.

Moderator. A material (in the RBMK1000 it is graphite) used in a reactor to slow down, or moderate, the fast neutrons produced by fission, thus increasing the likelihood of further fission.

Myelopoiesis. The formation of bone marrow or cells that arise from it.

Nano. Prefix for one-thousandth-millionth, i.e. 10^{-9}.

Neoplasm. Cancer.

Neutron. Elementary particle with a mass approximately equal to that of the proton, but with no electrical charge.

Noble gases. Chemically inert gases such as xenon and krypton.

NRPB. National Radiological Protection Board in the United Kingdom.

Nuclear accidents. *See* **Fermi reactor fuel melting, Kyshtym, Three Mile Island** and **Windscale.**

Nucleus of an atom. The central core of the atom, occupying only a small part of the atomic volume but containing most of the atomic mass. The nucleus of an atom contains all the positive charges, but the atom itself is electrically neutral since orbiting electrons around the nucleus contain negative charges, exactly equal, numerically, to the positive charges. The elementary particles in the nucleus are protons (these have the positive charges) and neutrons (these have no charge).

Penetration of alpha, beta and gamma rays. Early experiments with radioactive substances showed that they contained radiation components with different penetration powers. Alpha rays were the least penetrating and were completely absorbed by a few centimetres of air or by a very thin sheet of metal; beta rays were absorbed by about 1 millimetre of lead; and gamma rays, the most penetrating, were capable of passing through some 10 centimetres of lead.

Another difference was that only the alpha and beta rays could be deflected by a magnetic field. The nature of the three radiations was obviously different and it was shown that alpha rays are the nuclei of helium atoms, that is two protons plus two neutrons. Thus they are four times the mass and twice the positive charge of a proton. Beta rays are electrons and gamma rays are photons of electromagnetic radiation. The latter are not particles like the other two forms of radiation.

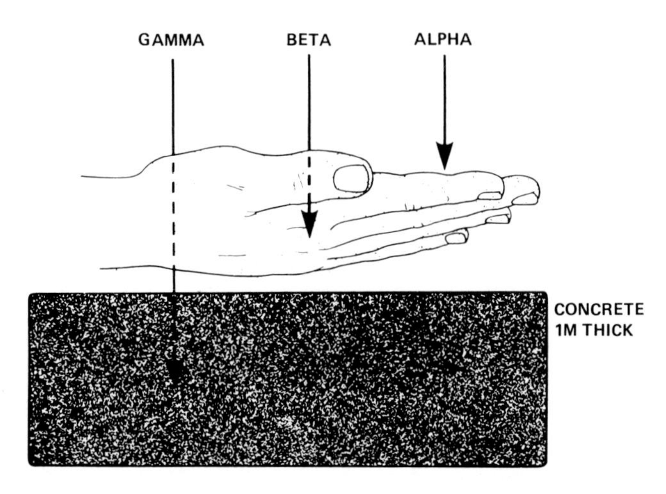

The penetrating power of radiation. Alpha rays will just penetrate the surface of the skin; beta rays can pass through 1–2 centimetres of human tissue; gamma rays are very penetrating and can pass right through the human body but would be almost completely absorbed by 1 metre of concrete.

Photon. *See* **Gamma ray** and **X-ray.**

Pigmentation. Deposition of colouring matter. The dark blue pigmentation of the skin experienced by the Chernobyl victims after beta-irradiation is melanin.

Plutonium. A radioactive metallic element which was discovered in 1940. It is formed when neutrons bombarded uranium-238. Plutonium-239 has a half-life of 24,100 years, emits alpha particles, and if bombarded by thermal neutrons is fissionable. This fission reaction of plutonium-238 was the basis for the atomic bomb at Nagasaki in 1945. Plutonium-238, with a half-life of 86 years, is another plutonium radioactive isotope. Plutonium is deposited in bones following intake by man.

Positive void coefficient. A nuclear reactor design, such as the RBMK1000, has what is known as a positive void coefficient if it contains the following design feature. If the amount of steam in the reactor core increases for any reason, then the power of the reactor tends to go up and still more water converts to steam which causes more power, and the process continues without limit. It is, though, possible to compensate for this undesirable characteristic by careful design. In the RBMK1000 this can only be achieved under normal operating conditions. However, at any reactor power below 20%, the reactor could have what is called a *positive power coefficient*. In this case a simple increase in power leads to a further increase in power and this escalates without stopping, as at Chernobyl.

Prognosis. Future prospect of recovery. Often used in relation to cancer patients.

Proton. Elementary particle with a positive charge equal and opposite to that of an electron.

PWR. Pressurised water reactor.

Rad unit. *See* **Absorbed dose** and **Radiation units: SI and non-SI: conversion factors.**

Radiation risk factor. The probability of cancer and leukaemia or hereditary damage per unit does equivalent. It usually refers to fatal cancers and serious hereditary damage. Currently, a risk is estimated at the lowest level of dose that gives observable results. On this basis 1 millisievert to the whole body corresponds to a risk of death of about 1 or 2 per 100,000 population – provided that the age at radiation exposure provides a long enough life expectation for cancers to occur. For each organ (e.g. thyroid, lung, bone, gonads) the risk per unit dose equivalent is smaller than it is for the whole body.

Radiation syndrome. *See* **Acute radiation syndrome.**

Radiation units: SI and non-SI: conversion factors. The 30th World Health Assembly in May 1977 endorsed the use of SI units in medicine.

Quantity	SI unit		Non-SI unit		Conversion factor
	Name	Symbol	Name	Symbol	
Radioactivity	becquerel	Bq	curie	Ci	$1\ Ci = 3.7 \times 10^{10}\ Bq$
Exposure	coulomb per kilogram	C/kg	rontgen	R	$1\ R = 2.58 \times 10^{-4}\ C/kg$
Absorbed dose	gray	Gy	rad	rad	1 rad = 0.01 Gy
Dose equivalent	sievert	Sv	rem	rem	1 rem = 0.01 Sv

Radioactive decay. The process of spontaneous disintegration of a radioactive isotope. This disintegration may be either by alpha decay or by beta decay.

Radionuclide. Alternative term for a radioactive isotope.

Radium. This radioactive isotope, radium-226, was discovered by Marie and Pierre Curie in 1898, and is a high-energy alpha particle emitter with a half-life of 1620 years. It is naturally occurring, and found in uranium ores such as pitchblende. For many years it was used in the form of needles and tubes (containing the sulphate or bromide mixed with an inactive filler and encapsulated in a platinum–iridium sheath) to treat cancer. However, because it decayed to a gas, radon-222, itself radioactive, and because of its very long half-life, it has now been largely replaced in medicine throughout the world by caesium-137, iridium-192, cobalt-60 and gold-198 radioactive isotopes.

RBE. Relative biological effectiveness.

RBMK1000. The type of reactors at Chernobyl nuclear power plant.

Reactivity. A measure of the ability of the fuel assembly in the core of a nuclear reactor to support a sustained chain reaction.

Rem unit. *See* **Dose equivalent** and **Radiation units: SI and non-SI: conversion factors.**

Rontgen unit. *See* **Exposure** and **Radiation units: SI and non-SI: conversion factors.**

Runaway. Accidentally uncontrolled chain reaction.

Scram. Emergency shutdown of the fission chain reaction in the reactor.

Sellafield. Current name of the Windscale site.

SI. Système International, as referring to units.

Sievert unit. *See* **Dose equivalent** and **Radiation units: SI and non-SI: conversion factors.**

Stochastic and non-stochastic effects. Effects for which the probability of the effect occurring, rather than the severity of the effect if it occurs, varies with the size of the radiation dose. For such effects, as in the induction of genetic defects or cancer, no threshold dose can be assumed below which some effect may not occur. Stochastic effects therefore refer to hereditary or carcinogenic effects of radiation. These contrast with the results of high radiation exposures (e.g. radiation skin burns) which are likely to occur whenever a threshold dose has been exceeded. These are commonly referred to as non-stochastic effects.

Strontium. The radioactive isotope strontium-90 is a fission product, which emits beta particles when it decays with a 28.6 year half-life.

Syndrome. A combination of symptoms resulting from a single cause or so commonly occurring together as to constitute a distinct clinical entity. *See also* **Acute radiation syndrome.**

Tera. Prefix for millions, i.e. 10^{12}.

Thermal burns. Heat burns. For the victims of the Chernobyl accident, thermal burns are distinguished from beta-radiation burns.

Thermal neutrons. Neutrons that have been slowed to the degree that they have the same average thermal energy as the atoms or molecules through which they are passing.

Three Mile Island. Site of a civil nuclear power plant accident in 1979 at Harrisburg, Pennsylvania, USA. The following description is taken from a 1 July 1986 Congressional Issue Brief, USA. "Without doubt, the most serious and publicised nuclear power accident in the United States occurred in March 1979 at unit 2 of the Three Mile Island Nuclear Station in Middletown, Pennsylvania, a pressurised-water reactor with a capacity of 906 MWe. Because of equipment and procedural failures, the core of the reactor became uncovered long enough for some fuel to melt. Radioactivity escaped into the containment building, which held most of it, but some got out to cause minor contamination near the plant. There was no full-scale evacuation, although the Governor of Pennsylvania did recommend that pregnant women leave the area, and tens of thousands of people fled the area on their own. Later, an inter-agency ad hoc Population Dose Assessment Group concluded that the off-site collective dose associated with the radioactive release represented minimal risks of additional health effects to the off-sited population. Cleanup of the damaged core is not yet completed."

Thrombocytes. A term for platelets. They play an important part in clotting the blood. *See also* **Leucocytes.**

Trip. Shutdown, when referred to a nuclear reactor.

UNSCEAR. United Nations Scientific Committee on the Effects of Atomic Radiation.

Uranium. Uranium is naturally occurring in minerals such as pitchblende. In 1896 the phenomenon of radioactivity was discovered by Henri Becquerel, using uranium salts. A uranium nucleus contains 92 protons and naturally occurring uranium is mainly uranium-238 with only 0.7% uranium-235. Only a fast neutron will cause fission of uranium-238, but thermal neutrons will cause fission of uranium-235.

WHO. World Health Organisation.

Windscale.* Site of a civil nuclear power plant accident in 1957 in the United Kingdom. The following description is taken from a 1 July 1986 Congressional Issue Brief, USA. "In the 1950s the British nuclear weapons program obtained some of its plutonium from a small graphite moderated air cooled production reactor at Windscale. During routine servicing in 1957 some of the graphite became so hot that some uranium fuel and some graphite caught fire. An estimated 20,000 curies of iodine-131 were released, which fell upon surrounding meadows and caused contamination of the milk supply. There was no evacuation of residents, although use of milk from most of the contaminated region was prohibited for 25 days, and from a few places for 44 days. A committee of Britain's Medical Research Council concluded that it was unlikely any harm had been done to the health of anyone, whether a worker at Windscale or a member of the general public."

Xenon poisoning of a nuclear reactor. Only a small proportion of xenon-135 is formed directly by fission. Most comes from radioactive decay of iodine-135, and the xenon-135 is removed partly by radioactive decay (half-life = 9.2 hours) and partly by its capture of neutrons. About 2% of all neutrons are captured by xenon-135 and therefore it is an important factor in the overall neutron balance in the reactor. The balance of formation of xenon-135 and its destruction are such that a fall in reactor power, and therefore of neutron flux, leads to a rise in xenon concentration. This is sometimes referred to as xenon poisoning when the xenon-135 concentration is too high.

X-ray. A discrete quantity of electromagnetic energy (sometimes called a *photon* of energy), without any mass or any electric charge. Emitted from an X-ray machine such as those used for diagnosis in medicine, or for treatment of cancer.

---

* See Additional Glossary, p. 241–243.

*References**

Some of these references will be useful for further reading and for identifying scientific work on the consequences of Chernobyl which have been carried out in various countries.

(Note: For references in the journal *Nature* it is not always obvious where the work was undertaken and therefore the country of origin is given in brackets at the end of the reference.)

ABRAMS, H. L., "How radiation victims suffer". *Bulletin of the Atomic Scientists*, Vol. 43, pp. 13–17, August/September 1986.

ALEXANDROPOULOS, N. G., ALEXANDROPOLOU, T., ANAGNOSTOPOULOS, D., EVANGELOU, E., KOTSIS, K. T. & THEODORIDOU, I., "Chernobyl fallout on Ionnina, Greece". *Nature*, Vol. 322, p.779, 28 August 1986. (Greece)

AOYAMA, M., HIROSE, K., SUZUKI, Y., INOUE, H. & SUGIMURA, Y., "High level radioactive nuclides in Japan in May". *Nature*, Vol. 321, pp. 819–20, 26 June 1986. (Japan)

APSIMON, H. & WILSON, J., "Tracking the cloud from Chernobyl". *New Scientist*, pp. 42–5, 17 July 1986.

ARAKI, T. & MOROTANI, Y. (Mayors of the cities of Hiroshima and Nagasaki), *Appeal to the Secretary General of the United Nations*. Including: I. Physical destruction due to the atomic bomb; II. Medical effects of the atomic bomb; III. Sociological destruction due to the atomic bomb; IV. Problems for future study. Illustrated with 49 photographs. October 1976.

BARABANOVA, A. V., BARANOV, A. E., GUSKOVA, A. K., KEIRIM-MARKUS, I. B., MOISEEV, A. A., PYATKIN, E. K., REDKIN, V. V. & SUVOROVA, L. A., *Acute radiation effects in man*. State Committee of the USSR on the Use of Nuclear Energy, National Commission on Radiation Protection at the USSR Ministry of Health. Moscow: TsNIIatominform-ON-3, 1986. (Translated from the Russian for Oak Ridge National Laboratory.)

BARRETT, A. J. & GORDON-SMITH, E. C., *Bone marrow transplantation: a review*. Oxford: Medicine Publishing Foundation, 1985.

BERRY, R. J., "Living with radiation – after Chernobyl". Editorial from *The Lancet*, pp. 609–10, 13 September 1986.

BERRY, R. J., "Chernobyl: the anatomy of a disaster". *Cancer Topics*, Vol. 6, pp. 40–2, 1987.

BLIX, H., *The influence of the accident at Chernobyl*. Report C8, Division of Public Information, Vienna: IAEA, 1986.

BLIX, H., *The post-Chernobyl outlook for nuclear power*. Address given to the European Nuclear Conference '86, Geneva, 2 June 1986, Vienna: IAEA, 1986.

BONDIETTI, E. A. & BRANTLEY, J. N., "Characteristics of Chernobyl radioactivity in Tennessee". *Nature*, Vol. 322, pp. 313–14, 24 July 1986. (USA)

BRITISH INSTITUTE OF RADIOLOGY, "Biological basis of radiological protection and its application to risk assessment". Proceedings of a one-day seminar held at the British Institute of Radiology 44th Annual Congress, 11 April 1986. Contents: UPTON, A. C., "Cancer induction and non-stochastic effects". LYON, M. F., "Hereditary effects of radiation: some evidence from animal experiments". MOLE, R. H., "Irradiation of the embryo and fetus". THORNE, M. C., "Principles of the ICRP system of dose limitation". BEAVER, P. F., "Practical implementation of ICRP recommendations". POCHIN, E. E., "Radiation risks in perspective". *British Journal of Radiology*, Vol. 60, pp. 1–50, 1987.

CASSEL, C. K., "Political and medical lessons of Chernobyl". *Journal of the American Medical Association*, Vol. 256, pp. 630–1, 1986.

* See Additional References, p. 243–244.

Central Electricity Generating Board, "Chernobyl special". *Power News*, September, 1986.

Clarke, R. H., "NRPB Response to Chernobyl" and "Chernobyl and the international agencies". *Radiological Protection Bulletin*, No. 75, August/September 1986, pp. 5–6 and pp. 10–12, NRPB, Chilton.

Collier, J. G. & Davies, L. M., *Chernobyl*. Report prepared for the Central Electricity Generating Board, UK, September 1986.

Darby, S. C., "Evaluation of radiation risk from epidemiological studies of populations exposed at high doses". *The Statistician*, Vol. 34, pp. 59–72, 1985.

Devell, L., Tovedal, H., Bergstrom, U., Appelgren, A., Chyssler, J. & Andersson, L., "Initial observations of fallout from the reactor accident at Chernobyl". *Nature*, Vol. 321, pp. 192–3, 15 May 1986. (Sweden)

Donnelly, W., Behrens, C., Martel, M., Civiak, R. & Dodge, C., *The Chernobyl nuclear accident: causes, initial effects, and congressional response.* Updated 1 July 1986, Issue brief, Order code IB86077, Environment & Natural Resources Policy Division & Science Policy Division, Congressional Research Service, USA.

Fry, F. A., Clarke, R. H. & O'Riordan, M. C., "Early estimates of UK radiation doses from the Chernobyl reactor". *Nature*, Vol. 321, pp. 193–5, 15 May 1986. (United Kingdom)

Geiger, H. J., "The accident at Chernobyl and the medical response". *Journal of the American Medical Association*, Vol. 256, pp. 609–12, 1986.

General, Municipal, Boilermakers & Allied Trades Union, *Report of the Union Delegation to the USSR*. Prepared by J. Edmonds, F. Alexander, F. Cottam, J. Norfolk, C. Roberts & others. Contents: Chapter 1, Reactor designs and characteristics; Chapter 2, Radiation; Chapter 3, The containment issue; Chapter 4, Employment and the nuclear industry; Chapter 5, The Chernobyl accident; Chapter 6, Report to the GMBATU Central Executive Council. London, 1987.

Gilbert, E. S., "How much can be learned from populations exposed to low levels of radiation?" *The Statistician*, Vol. 34, pp. 19–30, 1985.

Gittus, J. H, Hicks, D., Bonell, P. G., Clough, P. N., Dunbar, I. H., Egan, M. J., Hall, A. N., Nixon, W., Bulloch, R. S., Luckhurst, D. P. & Maccabee, A. R., *The Chernobyl accident and its consequences.* Report by members of the Safety & Reliability Directorate, UKAEA, Harwell Laboratory and the National Nuclear Corporation. United Kingdom Atomic Energy Authority, London: H.M. Stationery Office, 1987.

Gubaryev, V. (Science Correspondent of *Pravda*), *Sarcophagus.* Script of a play translated from the Russian by M. Glenny and used by the Royal Shakespeare Company in their April 1987 production at the Barbican Theatre, London. (Later published by Penguin Books, 1987.)

Guskova, A. K., "Basic principles of the treatment of local radiation injuries" in "Radiation damage to skin". Proceedings of a workshop held in Saclay, France, 9–11 October 1985. *British Journal of Radiology*, Supplement No. 19, pp. 122–5, 1986.

Guskova, A. S., *Early acute effects of the Chernobyl accident: acute radiation effects in victims of the accident at the Chernobyl nuclear power station.* Working document for 1987 meeting of the International Commission on Radiation Protection, ICRP/87/C:G-01.

Hawkes, N., Lean, G., Leigh, D., McKie, R., Pringle, P. & Wilson, A., *The worst accident in the world.* London: William Heinemann & Pan Books, 1986.

Haywood, J. K. (Editor), *Chernobyl: response of medical physics departments in the United Kingdom.* IPSM Report No. 50. London: Institute of Physical Sciences in Medicine, 1986.

Hohenemser, C., Deicher, M., Hofsass, H., Lindner, G., Recknagel, E. & Budnick, J. I., "Agricultural impact of Chernobyl – a warning". *Nature*, Vol. 321, p.817, 26 June 1986. (Federal Republic of Germany)

Holliday, B., Binns, K. C. & Stewart, S. P., "Monitoring Minsk and Kiev students after Chernobyl". *Nature*, Vol. 321, pp. 820–1, 26 June 1986. (United Kingdom)

Ilyin, L. A. (Editor), *Manual for the organisation of medical treatment of persons who have been affected by ionising radiation.* Ehergoatomizdat (Scientific Edition), Moscow, 1986. (Publishing approval given on 5 August 1986, print run of 4650 copies, Russian Language Edition.)

International Atomic Energy Agency, *Radiation – a fact of life.* Vienna: IAEA, 1979.

International Atomic Energy Agency, *Facts about low-level radiation.* Vienna: IAEA, 1979.

International Atomic Energy Agency, Press releases to November 1987, including those on the 25–29 August 1986 post-accident review meeting. Vienna: IAEA, 1986.

International Atomic Energy Agency, "Response to Chernobyl". *IAEA Bulletin*, Vol. 28, No. 2, pp. 61–5, 1986.

International Atomic Energy Agency, "Nuclear plant safety: response to Chernobyl". Miscellaneous papers, including those by A. Petrosyants and A. Salo, and including national reports from Sweden, Poland, Federal Republic of Germany, United Kingdom and USA. *IAEA Bulletin*, Vol. 26, No. 3, pp. 1–39, 1986.

International Atomic Energy Agency, "Delegation's trip reinforces USSR-IAEA co-operation". *IAEA Newsbriefs*, Vol. 2, No. 1, p. 1, 20 January 1987.

International Atomic Energy Agency, *One year after Chernobyl: the IAEA's actions and programmes in nuclear safety*. Report D3, Division of Public Information, Vienna: IAEA, June 1987.

International Commission on Radiological Protection, "Recommendations of the International Commission on Radiological Protection". *Annals of the ICRP*, Vol. 1, No. 3. Oxford: Pergamon Press, 1977.

International Nuclear Safety Advisory Group, "Summary report on the post-accident review meeting on the Chernobyl accident". *IAEA Safety Series*, No. 75, INSAG-1. Vienna, IAEA, 1986.

Izvestia, issues from 2 May 1986 to 31 December 1986.

Johanson, L. & Brynildsen, L., *The Chernobyl fallout – status for Norwegian meat production, April 1987*. Oslo: Ministry of Agriculture, Division of Veterinary Services, 1987.

Jones, R. R., "Cancer risk assessments in light of Chernobyl". *Nature*, Vol. 323, pp. 585–6, 16 October 1986. (United Kingdom)

Jost, D. T., Gaggeler, H. W., Baltensperger, U., Zinder, B. & Haller, P., "Chernobyl fallout in size-fractionated aerosol". *Nature*, Vol. 324, pp. 22–3, 6 November 1986. (Switzerland)

Kaul, A., *Chernobyl nuclear accident: quantification and assessment of risk from radiation*. Federal Republic of Germany: Federal Health Office, 1986.

Kazutin, D., "Forget Chernobyl?" Review of book manuscript by Andrei Illesh for Rybok Publishers, Sweden, in *Moscow News*, No. 46, p. 12, 1986.

Kelly, P., "How the USSR broke into the nuclear club". *New Scientist*, pp. 32–5, 8 May 1986.

Kereiakes, J. G., Saenger, E. L. & Thomas, S. R., "The reactor accident at Chernobyl: a nuclear medicine practitioner's perspective". *Seminars in Nuclear Medicine*, Vol. 16, pp. 224–30, 1986.

Ketchum, L. E., "Lessons of Chernobyl: Society of Nuclear Medicine members try to decontaminate world threatened by fallout". *Journal of Nuclear Medicine*, Vol. 28, pp. 933–42, 1987.

Ketchum, L. E., "Lessons of Chernobyl: health consequences of radiation released and hysteria unleashed". *Journal of Nuclear Medicine*, Vol. 28, pp. 413–22, 1987.

Kjelle, P. E., *Fallout in Sweden from Chernobyl: part 1*. SSI-rapport 86–20. Stockholm: Statens Stralskyddsinstitut, 1986.

Kretschmar, J. & Billiau, R. (Editors), *The Chernobyl accident and its impact*. Proceedings of a seminar of the Studiecentrum voor Kernergie/Centre d'etude de l'energie nucleaire held on 7 October 1986 at Mol, Belgium. Publication number 86.02. Mol: I.S. Publications, 1986.

Latarjet, R., "Sur l'accident nucleaire de Tchernobul". *Comptes Rendus Academie des Sciences*, Paris, Vol. 303, Series III, pp. 19–24, 1986.

Le Monde, *Les defis du nucleaire*. Special 16 page issue, February 1987, Paris.

Legasov, V., *The lessons of Chernobyl are important for all*. Moscow: Novosti Press Agency Publishing House, 1987. (In English)

Lindell, B. & Dobson, R. L., "Ionising radiation and health". *Public Health Papers*, No. 6, World Health Organisation, Geneva, 1961.

Lushbaugh, C. C., Fry, S. A. & Ricks, R. C., "Medical and radiobiological basis of radiation accident management". Presentation at the British Institute of Radiology seminar on "Nuclear reactor accidents: preparedness and medical consequences", Southampton, 2 April 1987. *British Journal of Radiology*, vol. 60, pp. 1159–63, 1987.

MacLeod, G. K. & Hendee, W. R., "Radiation accidents and the role of the physician: a post-Chernobyl perspective". *Journal of the American Medical Association*, Vol. 256, pp. 632–4, 1986.

Maddox, J., "Second chance for nuclear power? Soviet frankness creates sense of solidarity. Chronology of a catastrophe. Shutting the stable door. Tracking radiation release". Editorial in *Nature*, Vol. 323, pp. 1–3 and pp. 26–9, 4 September 1986.

Marshall, E., "Reactor explodes amid Soviet silence". *Science*, Vol. 232, pp. 814–15, 16 May 1986.

Marwick, C., "Physicians' reaction to Chernobyl explosion: lessons in radiation – and cooperation". *Journal of the American Medical Association*, Vol. 256, pp. 559–65, 1986.

McCally, M., "Hospital number six: a first-hand report". *Bulletin of the Atomic Scientists*, Vol. 43, pp. 10–12, August/September 1986.

Medvedev, Z., "Nuclear power? nyet, ta". *New Statesman*, pp. 18–19, 9 May 1986.

Merz, B., "No place to hide – computer models track atmospheric radionuclides worldwide". *Journal of the American Medical Association*, Vol. 256, pp. 566–7, 1986.

Morrey, M., Brown, J., Williams, J. A., Crick, M. J., Simmonds, J. R. & Hill, M. D., *A preliminary assessment of the radiological impact of the Chernobyl reactor accident on the population of the European Community*. United Kingdom: National Radiological Protection Board, January 1987. (Work funded under CEC contract number 86 398).

Morris, J. A., *Exposure of animals and their products to radiation – surveillance, monitoring, control of national and international trade*. Report presented at the 55th General Session of Office International des Epizooties, Paris, 18–22 May 1987. United Kingdom: Ministry of Agriculture, Fisheries and Food, 1987.

MOULD, R. F., "After Chernobyl". *British Institute of Radiology Bulletin*, June 1987, pp. B29–B34.

MOULD, R. F., *Cancer Statistics*. Includes chapters on cancer registries, cancer incidence, cancer risk, cancer prevalence, cancer mortality, cancer survival, cancer treatment success and cancer cure. Bristol: Adam Hilger, 1983.

MOULD, R. F., *Radiation protection in hospitals*. Includes chapters on atoms, radioactivity and X-rays; radiation risk; radiation absorption and attenuation; radiation measurement; radiation shielding; classification of radiation workers; protection in: external beam radiotherapy, interstitial source radiotherapy, intracavitary radiotherapy, radioactive iodine-131 radiotherapy, nuclear medicine radiodiagnostics, and diagnostic radiology. Bristol: Adam Hilger Ltd, 1985.

MOULD, R. F., *A history of X-rays and radium*. Chapters include some early medical investigations; early treatment techniques; and radiation units 1895–1937. Sutton: IPC Building & Contract Journals Ltd, 1980.

NAPALKOV, N. P., TSERKOVNY, G. F., MERABISHVILI, V. M., PARKIN, D. M., SMANS, S. & MUIR, C. S. (Editors), *Cancer incidence in the USSR*, IARC Scientific Publication No. 48, 2nd revised edition. Lyon: International Agency for Research on Cancer, 1983.

NATIONAL INSTITUTE OF RADIATION PROTECTION, "After Chernobyl? Implications of the Chernobyl accident for Sweden". Special issue of *News & Views: Information for immigrants*, Stockholm, November 1986.

NATIONAL RADIOLOGICAL PROTECTION BOARD, *Living with radiation*, 3rd edition. Chilton: NRPB, 1986.

NICHOLSON, R. A., NICHOLSON, J. P. & MOULD, R. F., "Westminster Hospital monitoring with a single sodium iodide counter". In *Chernobyl: response of medical physics departments in the United Kingdom*, edited by J. K. Haywood, pp. 27–32. Institute of Physical Sciences in Medicine, London, 1986.

NISHIZAWA, K., TAKATA, K., HAMADA, N., OGATA, Y., KOJIMA, S., YAMASHITA, O., OHSHIMA, M. & KAYAMA, Y. "Iodine-131 in milk and rain after Chernobyl". *Nature*, Vol. 324, p. 308, 27 November 1986. (Japan)

NOVOSTI PRESS AGENCY, "Novosti Press Agency reports on Chernobyl", May 1986–November 1987.

OFFICE INTERNATIONAL DES EPIZOOTIES, *Control of radioactivity in man's food and in animals*. Report of the 54th General Session of the O.I.E., 26–30 May 1986. Report number 54 SG/RF, p. 22, paragraph 127. Paris: O.I.E., 1986.

ORLANDO, P., GALLELLI, G., PERDELLI, F., DE FLORA, S. & MALCONTENTI, R., "Alimentary restrictions and iodine-131 in human thyroids". *Nature*, Vol. 324, p. 23, 6 November 1986. (Italy)

PATTERSON, W. C., *Nuclear power*, 2nd edition. London: Penguin Books, 1986.

PERMANENT MISSION OF THE SOVIET UNION, GENEVA, *Press bulletins on Chernobyl*, 11 May–27 July 1986.

PETROSYANTS, A., "The Soviet Union and the development of nuclear power". *IAEA Bulletin*, Vol. 28, No. 3, pp. 5–8, 1986.

PETROSYANTS, A. M., "Obninsk marks 30 years of nuclear power". *IAEA Bulletin*, Vol. 26, No. 4, pp. 42–6, 1984.

POCHIN, E., *Nuclear radiation: risks and benefits*, Oxford: Clarendon Press, 1983.

POHL, F., *Chernobyl – a novel*. London: Bantam Books, 1987.

POURCHET, M., PINGLOT, J. F. & GASCARD, J. C., "The northerly extent of Chernobyl contamination". *Nature*, Vol. 323, p. 676, 23 October 1986. (France)

Pravda, issues from 2 May 1986 to 25 December 1986.

PRINGLE, D. M., VERMEER, W. J. & ALLEN, K. W., "Gamma-ray spectrum of Chernobyl fallout". *Nature*, Vol. 321, p. 569, 26 June 1986. (United Kingdom)

RASSOW, J., *Kernreaktorunfall in Tschernobyl*. Federal Republic of Germany: Universitatsklinikum der Universitat-Gesamthochschule-Essen, 1987.

SADASIVAN, S. & MISHRA, U. C., "Relative fallout swipe samples from Chernobyl". *Nature*, Vol. 324, pp. 23–4, 6 November 1986. (India)

SAENGER, E. L., "Radiation accidents". *Annals of Emergency Medicine*, Vol. 15, pp. 1061–66, 1986.

SALO, A., "Information exchange after Chernobyl". *IAEA Bulletin*, Vol. 28, No. 3, pp. 18–22, 1986.

SCHAFER, H., *Endlager-statte Mensch?* Munich: Knauer, 1986.

SCHERBAK, Y., "Chernobyl: documentary story". *Unost* (Youth), June 1987, pp. 46–66. (In Russian)

SEMENOV, B. A., "Nuclear power in the Soviet Union". *IAEA Bulletin*, Vol. 25, No. 2, pp. 47–59, 1983.

STATENS STRALSKYDDSINSTITUT, *Chernobyl – its impact on Sweden*. SSI-rapport 86–12. Stockholm: National Institute of Radiation Protection, 1986.

The Biologist, "The lessons of Chernobyl". Proceedings of the 11 April 1987 Institute of Biology

Seminar in London which included contributions from F. R. Livens, T. Hugosson, H. D. Roedler, A. Guskova and R. Gale.

THOMAS, A. J. & MARTIN, J. M., "First assessment of Chernobyl radioactive plume over Paris". *Nature*, Vol. 321, pp. 817–19, 26 June 1986. (France)

TRADES UNION CONGRESS NUCLEAR ENERGY REVIEW BODY, *Report of a delegation visit to the USSR, hosted by the All Union Central Council of Trade Unions, 2–5 April 1987*. London: Trades Union Congress, 1987.

UNION FEDERALE DES CONSOMMATEURS, "Tchernobyl ce qui est reste radioactif". *Que Choisir?*, special issue, 1987.

UNITED NATIONS ENVIRONMENT PROGRAMME, *Radiation doses, effects, risks*. Geneva: UNEP, 1985.

UNITED NATIONS SCIENTIFIC COMMITTEE ON THE EFFECTS OF ATOMIC RADIATION, *Exposures resulting from nuclear weapons test explosions and the military fuel cycle*. Proceedings of the 35th Session of UNSCEAR, Vienna, 14–18 April 1986. Vienna: UNSCEAR, 1986.

UNITED STATES FOOD AND DRUG ADMINISTRATION, FDA Press Office daily clipping service: Chernobyl, 29 April–13 May 1986, Rockville, Maryland.

UNITED STATES NUCLEAR REGULATORY COMMISSION, *Report on the accident at the Chernobyl nuclear power station*. NUREG-1250. Prepared by: Department of Energy, Electric Power Research Institute, Environmental Protection Agency, Institute of Nuclear Power Operations and Nuclear Regulatory Commission, Washington DC., 1986.

USSR STATE COMMITTEE ON THE UTILIZATION OF ATOMIC ENERGY, *The accident at the Chernobyl nuclear power plant and its consequences*. Information compiled for the IAEA Experts' Meeting, 25–29 August 1986. Vienna: IAEA, 1986.

VAN DER VEEN, J., VAN DER WIJK, A., MOOK, W. G. & DE MEIJER, R. J., "Core fragments in Chernobyl fallout". *Nature*, Vol. 323, pp. 399–400, 2 October 1986. (The Netherlands)

VARLEY, J., "Chernobyl prepares to start up". Editorial in *Nuclear Engineering International*, Vol. 31, p. 7, 1986.

VOLCHOK, H. L. & CHIECO, N. (Editors), *A compendium of the Environment Measurements Laboratory's research projects related to the Chernobyl nuclear accident*. EML-460. New York: US Department of Energy, 1986.

VON HIPPEL, F. & COCHRAN, T. B., "Estimating long-term health effects". *Bulletin of the Atomic Scientists*, Vol. 43, pp. 18–24, August/September 1986.

WAIGHT, P. J., "Estimate of the relationship between food contamination and consumption". Personal communication, 31 March 1987.

WATERHOUSE, J. A. H., MUIR, C. S., CORREA, P. & POWELL, J. (Editors), *Cancer incidence in five continents*, Volume III. IARC Scientific Publication, No. 15. Lyon: International Agency for Research on Cancer, 1976.

WATERHOUSE, J. A. H., MUIR, C. S., SHANMUGARATNAM, K. & POWELL, J. (Editors), *Cancer incidence in five continents*, Volume IV. IARC Scientific Publication, No. 42. Lyon: International Agency for Research on Cancer, 1982.

WATSON, W. S., NICHOLSON, R. A. & MOULD, R. F. "Chernobyl radionuclide deposition in Kiev and Warsaw". *Nature*, in press, 1988. (United Kingdom)

WEBB, G. A. M., SIMMONDS, J. R. & WILKINS, B. T., "Radiation levels in Eastern Europe". *Nature*, Vol. 321, pp. 821–2, 26 June 1986. (United Kingdom)

WEBSTER, E. W., "Chernobyl predictions and the Chinese contribution". *Journal of Nuclear Medicine*, Vol. 28, pp. 423–5, 1987.

WEINBERG, S., "Armand Hammer's unique diplomacy". *Bulletin of the Atomic Scientists*, Vol. 43, pp. 50–2, August/September 1986.

WORLD HEALTH ORGANISATION REGIONAL OFFICE FOR EUROPE, *Chernobyl reactor accident*. Report of a consultation. Copenhagen: WHO, 1986.

WORLD HEALTH ORGANISATION REGIONAL OFFICE FOR EUROPE, *Assessment of radiation dose commitment in Europe due to the Chernobyl accident*. Unpublished working group report, 25–27 June 1986. Bilthoven: WHO, 1986.

WORLD HEALTH ORGANISATION REGIONAL OFFICE FOR EUROPE, *Updated background information on the nuclear reactor accident in Chernobyl, USSR*. Data summary with regard to activity measurements, 12 June 1986.

SELECTED REFERENCES FOR FURTHER READING

This book concerns the Chernobyl nuclear accident, but readers may also be interested in references to the previous two worst civil nuclear accidents at Windscale in the United Kingdom in 1957 and at Three Mile Island, Pennsylvania, in the USA in 1979.

Windscale

Atomic Energy Office, *Accident at Windscale No. 1 pile on 10 October 1957.* Report of the Committee of Inquiry. Cmnd. 302. London: H.M. Stationery Office, 1957.

Mayneord, W. V., Anderson, W., Bentley, R. E., Burton, L. K., Crookall, J. O. & Trott, N. G., "Radioactivity due to fission products in biological material". *Nature*, Vol. 182, pp. 1473–8, 29 November 1958.

Thompson, T. J. & Beckerley, J. G. (Editors), *The technology of nuclear reactor safety – Volume 1: Reactor physics and control.* Chapter 11, "Accidents and destructive tests", Section 3.7, "Accident at Windscale No. 1 pile", pp. 633–6. Cambridge, Massachusetts: The M.I.T. Press.

Three Mile Island

Cantelon, P. L. & Williams, R. C., *Crisis contained, the Department of Energy at Three Mile Island: a history*, Chapter 1, "Accident at Three Mile Island", pp. 1–7. Washington DC: US Department of Energy, 1980.

Moss, T. H. & Sills, D. L. (Editors), "The Three Mile Island nuclear accident: lessons and implications". *Annals of the New York Academy of Sciences*, Vol. 365, 24 April 1981.

Sills, D. L., Wolf, C. O. & Shelanski, V. B. (Editors), *Accident at Three Mile Island: the human dimension.* Boulder, Colorado: Westview Press, 1982.

Stephens, M., *Three Mile Island.* London: Junction Books, 1980.

Hiroshima

These references which include eye witness accounts of some of the victims of Hiroshima as distinct from other more extensive scientific and technical studies. The latter are issued in the Life Span Study reports of the Japanese Ministry of Health & Welfare in co-operation with the Atomic Bomb Casualty Commission (replaced by the Radiation Effects Research Foundation from April 1975).

Chisholm, A., *Faces of Hiroshima.* London: Jonathan Cape, 1985.

Hersey, J., *Hiroshima.* London: Penguin Books, 1984.

January 1988 Update

CHAPTER 1 (pp. 1–3)

Note 1, p. 1
During my 2 December 1987 visit to Chernobyl I saw the outer walls of the buildings for units Nos. 5 and 6 apparently nearly completed, although still surrounded by cranes.

CHAPTER 2 (pp. 5–6)

Note 2, p. 6
During my 2 December 1987 visit to Chernobyl this was remarked upon with some amusement and a statement made that US satellites were not as good as Russians had previously thought!

CHAPTER 3 (pp 7–21)

Note 3, p. 18
A new director, M. P. Umanets, was appointed in February 1987.

Note 4, p. 18
Minister N. Lukonin of the Ministry of Atomic Energy is now (2 December 1987) responsible for the Chernobyl power station.

CHAPTER 5 (pp. 73–76)

Note 5, p. 76
During the author's 1–6 December 1987 visit to Chernobyl, Pripyat and Moscow, Mr Yuri Kanin of the Novosti Press Agency presented his 12-page article "*Chernobyl: the Evacuation*" which included the following comments on evacuation from villages. "It proved more difficult to evacuate people from villages (i.e.

compared with the town of Pripyat) . . . not sensing any immediate danger they did not see any special need to leave their organised way of life and their property. Many simply could not be convinced to leave their village before the problem of what to do with the animals (cows and pigs) was solved. . . . In most villages livestock had to be evacuated first, and only then people."

CHAPTER 6 (pp. 113–120)

Note 6, p. 117
The sleeping cars have very narrow corridors and are either hard class (4 bunks/compartment) or soft class (2 bunks/compartment) with limited toilet facilities, no restaurant car, no space for any large cases and next to no room to move in any compartment. On the nights of 1–2 and 2–3 December 1987 the express was full for the 12-hour journey, so in May 1986 the journey must have been uncomfortable in the extreme with compartments and corridors totally jammed with people.

CHAPTER 7 (pp. 121–131)

Note 7, p. 131
At the end of 1987 agreement between European countries was no nearer. For example, in *The Times* of 9 November 1987 it was reported that on the previous day the European Community Deputy Foreign Ministers had failed to agree on levels of permitted radiation in food that would apply in the event of a nuclear accident. Also, all countries except Greece did agree to enforcing standards introduced shortly after the Chernobyl accident, until 24 November – although they have officially expired. These standards had included an EEC ban (from 12 May to 30 May 1986) on all imports of fresh food and livestock from within a 1000-kilometre radius of Chernobyl. In the United Kingdom, the Ministry of Agriculture, Fisheries & Food set intervention levels of 1000 Bq/kg (in Sweden it was 300 and in Federal Republic of Germany, 600) total caesium activity. A ban was imposed on the movement and slaughter of animals exceeding this level and a guarantee was given that meat exceeding this level would not be marketed. Initially, in June 1986, this involved some 4.2 million sheep but by August 1987 the number under restriction had been reduced to 500,000, distributed among 564 hill farms. By 28 October 1987 U.K. Government compensation to some 7500 farmers had been of the order of £4.5 million.

The discussion concerning intervention levels will probably not end for some time. Not least because the group of experts set up under article 31 of the Euratom treaty have now recommended a level of 5000 Bq/kg, but this has not been accepted by the European Commission which is recommending a 75% reduction in the level proposed by their own experts.

Product	Activity (Bq/kg) limits for different radioactive isotopes			
	Caesium	Iodine	Strontium	Plutonium
Dairy products	1000	500	500	20
Foodstuffs other than dairy products	1250	3000	3000	80
Drinking water	800	400	400	10
Animal feedstocks	2500			

European Commission recommendations which are being considered for new limits to replace the emergency limits which expired on 31 October 1987 (After Johnston, K., *Nature*, 20 August 1987)

CHAPTER 8 (pp. 133–136)

Note 8, p. 134

On 2 December 1987 the temperature inside the sarcophagus was said to be 82°C, and also that it now housed 300 monitors for measurement of radiation, temperature, etc.

Note 9, p. 135

On 2 December 1987 filter systems were obvious, on windows in the administration building at the power station, in the government commission buildings in Chernobyl town, and at Zelyony Mys, just outside the 30-kilometre zone.

Note 10, p. 136

I have also heard it said that in order to complete the sarcophagus as soon as possible workers were seconded from every major building project in the Soviet Union. Whatever the truth of this remark, there is no doubt that volunteer workers came to Chernobyl from many parts of the Soviet Union to complete a most impressive feat of construction.

Note 11, p. 171

The following poster (shown bottom left in Fig. 153) was still displayed when the author visited Chernobyl on 2 December 1987, by which time the central banner had been removed.

CHAPTER 9 (pp. 179– 192)

Note 12, p. 181

The following three photographs were taken in February 1987 but only received by TASS in London in December 1987. The sign on the left of the first photograph translates as "Point. Dosimetry. Control Number 1". On 2 December 1987, those manning the dosimetry control points were still wearing face masks. In the next

STOP
ПУНКТ
ДОЗИМЕТРИЧЕСКОГО
КОНТРОЛЯ № 1

ОБЬЕЗД
ЧЕРНОБЫЛЬ
ПРИПЯТЬ

photograph the main road sign directs traffic straight ahead for Chernobyl and to the right for Pripyat. The driver is being spoken to by two members of the State Traffic Militia and the sign on the left states "Diversion". In the final photograph the sign drawn in Fig. 5 from *Izvestia* is illustrated. The debris piled on the verges has now been cleared away. It was bulldozed there when the shoulders of the road were asphalted.

Note 13, p. 188
A radiobiology laboratory had been established by 2 December 1987.

GLOSSARY (pp. 215–226)

Additional Glossary

Aspermia. Absence of seminal fluid or spermatozoa.

Atrophy. Decrease in size of organs or tissue.

Erythema of the skin. Redness of the skin. In the early years following the discoveries of X-rays and of radium, *skin erythema doses* were used in medical practice to specify the dose prescription with either X-rays or with radium. However, some of the patients were doubtful about the effect of radiation and a Viennese doctor recorded in 1901 that "An occasional slight erythema is the sole visible result of the treatment, so that patients are apt to become very sceptical of its success." Self-exposure experiments, usually on the hand or forearm, were a common procedure by practitioners who used the erythema dose to determine X-ray patient exposure and as late as 1927 a textbook even recommended that "because of medico-legal complications it is better for the operator to use his own skin". This was in spite of an alternative procedure described in 1903 at a meeting of The Röntgen Society in London: "Almost every worker with X-rays has to use the flourescent screen very frequently in order to test the condition of his X-ray tube by reference to a shadow of his own hand. This process has so often to be gone through that most workers have had to suffer through more or less acute inflammation of the skin of their hands. To obviate this is the function of the *Chiroscope*. This instrument consists of an articulated skeleton hand suitably mounted behind a small flourescent screen, which is to serve as a test object. The fleshy parts of the hand are to be represented by suitably cut out tin foil, and the whole is mounted on a holder, the construction of which affords protection to the hand of the operator."

Goiania, Brazil, radiation accident. Two incidents involving the theft of radioactive material from radiotherapy machines intended for scrap have occurred in the 1980s. The first was in Juarez, Mexico, in 1983, and the most recent, in September 1987, was in Brazil. The following text is reproduced from an IAEA press release (15 October 1987, PR 87/35).

> "Further information has now been made available to the IAEA from Brazil about the origin and consequences of the radiation incident in Goiania, capital city of the State of Goias.
>
> According to the information received, the incident followed the theft of a disused caesium-137 source which had been used for medical treatment at the local radiotherapy institute. Although it had not been used for some time, the source had been stored in a closed bunker. The thieves sold the source itself, with its protective shielding, to a scrap metal dealer who, not realising that the material he was handling was radioactive, broke open the container. The scrap metal dealer, his family, and some other persons who visited his premises, became contaminated.

> Within a few hours, these persons developed symptoms characteristic of over-exposure to radiation and went to the local hospital for treatment. It was at this stage that the incident was detected, and the national atomic energy commission was notified.
>
> More than 40 Brazilian experts were sent immediately to Goiania. They initiated procedures to define the affected area, and to monitor additional persons who might have been contaminated. The persons who were found to have been most seriously contaminated were sent to a naval hospital in Rio de Janeiro, where appropriate facilities for their treatment are available. Other, less seriously contaminated, persons were kept in hospital in Goiania. Seven contaminated areas were identified and isolated, and are now being decontaminated.
>
> The IAEA received from Brazil on 6 October a request for assistance in accordance with the terms of the Convention on Assistance in the Case of a Nuclear Accident or Radiological Emergency, to which Brazil is a signatory. At present, assistance is being rendered by experts from Argentina, the Federal Republic of Germany, the Soviet Union and the United States.
>
> According to the Brazilian authorities, the situation in Goiania is now considered to have been brought under control. However, at least four of the contaminated persons are in a critical condition.
>
> The Brazilian Government has announced an official inquiry into the incident."

Three or four persons have now died, including the 6-year-old daughter of the scrap dealer, who was reported in the media to have covered herself with the caesium-137 powder and even to have eaten some. Television news programmes showed one of the victims being buried in a lead-lined coffin and a contaminated dog being left to die in pain. According to *Time* of 19 October 1987, the Brazilian authorities have monitored more than 4000 people and 30 families have been evacuated from their homes and about 20 people have been hospitalised. The *Financial Times* of 14 October 1987 stated that 243 people were contaminated, and the *New Scientist* of 15 October 1987 that the activity of the radioactive source was 1400 curies of caesium-137.

IARC. International Association for Cancer Research, Lyon.

Juarez, Mexico, radiation accident. The equipment involved was a Picker Corporation cobalt-60 radiotherapy machine, manufactured some time before 1963 with the most recent supply of cobalt-60 sources in 1969. This activity was initially some 3000 curies of cobalt-60, which has a half-life of 5.3 years. The accident occurred in December 1983.

Unlike the more modern cobalt-60 radiotherapy machines which have all the cobalt-60 concentrated into a single small disc or rod with maximum dimensions of some 2 cm, the Picker source consisted of about 7000 tiny pellets, each of 1 mm diameter.

The machine had originally been sold to the Methodist Hospital in Lubbock, Texas, who, when it was no longer required, sold it to the X-ray Equipment Company of Fort Worth who in 1977 shipped the machine to the Centro Medico in Juarez. It was never in fact ever installed in the Centro Medico's radiotherapy department and remained stored in a warehouse until November 1983.

Someone then decided to dismantle it and it was stolen, eventually arriving in a pick-up truck at the Junke Fenix, a Juarez scrap metal yard, on 6 December 1983.

The whole sequence of events became known a month later on 16 January 1984, but *this was only by chance.*

A truck loaded with steel rods made from scrap took a wrong turning at the Los Alamos National Laboratory in New Mexico, USA, and happened to pass over a radiation sensor in the road outside the laboratory – and set off an alarm.

In the month after the break-up of the machine, two steel foundries in Mexico and one in the USA handled radioactive steel. Half of the contaminated metal went to a mill in Aceros, 160 Km south of Juarez, where reinforcing rods for concrete construction were made and officials found cobalt-60 in the clean-up water at Aceros de Chihuahua – showing at least some of the cobalt-60 had become airborne.

Many of the rods never left Aceros, but some of the radioactive steel from the Falcon de Juarez foundry was shipped to the US plant of Falcon Products, St. Louis, Missouri, which manufactures steel for the legs of restaurant table stands.

It was not until 25 January 1984 that the movement of steel was stopped and in that time some 5000 tonnes of reinforcing rods and some 18,000 table legs had left Mexico.

The highly contaminated pick-up truck had been parked for a month in a housing neighbourhood in Juarex and it was reported that at least 12 children had played on the truck before it was removed to safety. Some of the cobalt-60 pellets were found in the pick-up truck.

Oedema. Excess of water in the body.

Pedicle. A footlike or stemlike part, or attaching structure.

Polonium. Discovered by Pierre and Marie Curie in 1898, before their discovery of radium. Polonium has an atomic number of 84 and isotopes with mass numbers of 218, 216, 215, 214, 212, 211 and 210.

Telangiectasia. Presence of small red focal lesions, usually in skin or mucous membrane.

Wigner effect. Displacement of carbon atoms in the crystal lattice of graphite as a result of continued bombardment with neutrons. It leads to changes in overall shape and size of graphite, and the build-up of stored or potential energy which may be subsequently released as heat. The effect is serious only in reactors operating at low temperatures, below those needed for power production. (From UKAEA Press Release 28/87.)

Windscale. The Windscale accident of 1957 became headline news again in January 1988 because of the rule that the United Kingdom Cabinet papers cannot be released to the public until 30 years have elapsed. The newly available documents revealed that the then Prime Minister, Harold Macmillan, vetoed full publication of Sir William Penny's report and the Government only published a summary of the findings. This was the Atomic Energy Office November 1957 white paper *Accident at Windscale No. 1 Pile on 10th October 1957*. Typical news commentaries on the censoring of information for 30 years included:

From The Times *of 1 January 1988*

> A quotation from a 1957 Atomic Energy Authority minute "even if it had been considered that there was no security objection to the publication of so much technical detail (the then chief scientist at the Ministry of Defence advised the Prime Minister that there were 'no security objections' to releasing the full conclusions) there would still remain the danger that it would be quoted out of context and misused in other ways by hostile critics. In particular, it would provide ammunition to those in the United States who would in any case oppose the amendments of the McMahon Act (the 1954 law forbidding the American Government from sharing nuclear information with other countries) which the United States authorities intended to propose in order to make possible the desired degree of collaboration."

From the New Scientist *of 7 January 1988*

> A quotation from a member of the Atomic Energy Authority in 1957, "publication of the report would severely shake public confidence in the authority's competence to undertake the tasks entrusted to them and would inevitably provide ammunition for all those who had doubts of one or another about the development and future of nuclear power."

This journal also referred to the lack of mention of the alpha emitting radioactive isotope polonium-210 in the Penney report, although to be fair, the *New Scientist* suggested that this mistake "may have been because monitoring initially concentrated on emitters of gamma radiation". Nevertheless,

> "polonium-210 was being produced by the irradiation of bismuth in one of a number of side channels in the pile at the time of the accident. Papers written later by Atomic Energy Authority staff but not declassified until the 1980s reveal widespread polonium contamination. But neither the Penny report, nor any other of the documents released in January 1988, mention it. And the National Radiological Protection Board was not told about it when it first reviewed the Windscale data in 1982."

From the Daily Telegraph *of 2 January 1988*

> which interviewed Professor John Kay, a member of the four man team (with Penney, Professor Jack Diamond and Dr B. F. J. Schonland) which investigated the Windscale accident, quoted him as saying in 1988: "I would not wish to sound complacent, but certainly, if we had to have an accident, it is better that it happened when it did. It made everyone very safety concious and we got a good safety control set up in our nuclear industry. I crtainly feel very confident about the standards of safety today. We have good reactors, good management and good inspection."

The United Kingdom Atomic Energy Authority also issued press statements, both on the anniversary of the accident (Release 28/87, 5 October 1987, *Safety Lessons from Race to Build Britain's A-Bomb*, and Release 29/87, 5 October 1987, *Clean-up Starts on Windscale Piles*) and on January 1 1988 (Release 40/87, *Papers on 1957 Nuclear Fire Released*). In addition, the UKAEA journal *Atom* in October 1987 issue number 372, published *The Windscale fire 1957: a bibliography (of some 70 reports) of publicly available material* and in December 1987 issue number 374, published *The Windscale piles – past, present and future*. This

> "recounts the events leading up to the accident, the subsequent actions and surveillance undertaken to ensure their safety, and the work that is being done to prepare for decommissioning."

From the UKAEA Press Release *29/87*

> "Over the years the Piles have not posed any risk to workers on site or to the public. In fact for many years we have been using the space left from the earlier stages of clean-up as laboratories containing other large research rigs, and for offices."
>
> And
>
> "The clean-up operations will probably take about 10 years (from 1988). It is necessary to proceed cautiously and before any operations are undertaken on the reactors themselves, the (400-foot high) chimneys will be sealed by air dams at the base of each unit. We shall not know exactly what the state of the damaged core is until the air dams are in place and we can examine it more closely."

The fuel storage pond, which lies between the two reactors (Piles Nos. 1 and 2), will be emptied and cleaned up by British Nuclear Fuels Ltd, who will also undertake the work on the chimnies. The UKAEA will be responsible for decommissioning the reactors.

From the UKAEA Pree Release *40/87*

"The military Pile used for producing weapons material and not electricity, was the last reactor to be built (in the UK) without any form of containment. The cause of the fire was overheating of the Piles through the build-up of 'Wigner' energy, a phenomenon which caused the graphite blocks in the reactor's core to catch fire. This cannot occur in present day nuclear reactors."

and

"We welcome the disclosure of the files as an important addition to the nuclear debate and an indication to people of the many improvements to nuclear safety made since the 1950's."

REFERENCES (pp. 227–232)

Additional References

Berry, R. J., "The International Commission on Radiological Protection – a historical perspective". In *Radiation and Health*, edited by R. R. Jones & R. Southwood, Chapter 10, John Wiley & Sons, Chichester, 1987.

Bourdillon, P. J., "The role of National Health Service hospitals in the preparedness for nuclear accidents". *British Journal of Radiology*, Vol. 60, pp. 1171–4, 1987.

Clarke, R. H., "Reactor accidents in perspective". *British Journal of Radiology*, Vol. 60, pp. 1182–8, 1987.

Clarke, R. H., "Dose distributions in Western Europe following Chernobyl". In *Radiation and Health*, edited by R. R. Jones & R. Southwood, Chapter 20, John Wiley & Sons, Chichester, 1987.

Cox, R. A. F., "Nuclear emergencies: medical preparedness". *British Journal of Radiology*, Vol. 60, pp. 1180–2, 1987.

Csongor, E., Kiss, A. Z., Nyako, B. M. & Somorjai, E., "Chernobyl fallout in Debrecen, Hungary". *Nature*, Vol. 324, 20 November 1986. (Hungary).

Edmondson, B., "United Kingdom nuclear reactor design and operation". *British Journal of Radiology*, Vol. 60, pp. 1174–7, 1987.

Ennis, J. R., "New dosimetry at Hiroshima & Nagasaki – implications for risk estimates". Report of 23rd. Annual Meeting of the National Council on Radiation Protection & Measurements, Washington, DC, 8–9 April 1987, *Radiological Protection Bulletin*, No. 85, pp. 24–7, September 1987.

Fowler, S. W., Buat-Menard, P., Yokoyama, Y., Ballestra, S., Holm, E. & Nguyen, H. V., "Rapid removal of Chernobyl fallout from Mediterranean surface water by biological activity". *Nature*, Vol. 329, pp. 56–8, 3 September 1987. (Monaco & France)

Fry, F. A., "The Chernobyl reactor accident: the impact on the United Kingdom". *British Journal of Radiology*, Vol. 60, pp. 1147–58, 1987.

Gubaryev, V., "In Chernobyl's Sarcophagus". *The Scientist*, p. 23, 21 September 1987.

Hamman, H. & Parrott, S., *Mayday at Chernobyl*. Sevenoaks: New English Library, 1987.

House of Commons Official Report, Parliamentary Debates (Hansard), "Foodstuffs (Radioactive Contamination)" (Col. 403), Vol. 121, No. 33, London: HMSO, 29 October 1987.

Johnston, K., "British sheep still contaminated by Chernobyl fallout". *Nature*, Vol. 328, p. 661, 20 August 1987. (United Kingdom)

Johnston, K., "United Kingdom upland grazing still contaminated". *Nature*, Vol. 326, p. 821, 30 April 1987. (United Kingdom)

Jones, R. R. & Southwood, R., Editors, *Radiation and Health, the Biological Effects of Low-Level Exposure to Ionizing Radiation*. Chichester: John Wiley & Sons, 1987.

Lambert, B. E., "The effects of Chernobyl". In *Radiation and Health*, edited by R. R. Jones & R. Southwood, Chapter 21, John Wiley & Sons, Chichester, 1987.

Marples, D. R., *Chernobyl and Nuclear Power in the USSR*. Basingstoke: Macmillan, 1987.

Nadezhina, N. M., "Experience of a specialised centre in the organisation of medical care of persons exposed during a nuclear reactor accident". *British Journal of Radiology*, Vol. 60, pp. 1169–70, 1987.

Nationl Radiological Protection Board, "Interim guidance on the implications of recent revisions of risk estimates and the ICRP 1987 Como statement". NRPB–GS9, November 1987.

National Radiological Protection Board, "Statement from the 1987 Como meeting of the International Commission on Radiological Protection". Supplement to *Radiological Protection Bulletin*, No. 86, 1987.

NEFFE, J., "Germany and Chernobyl, end of the nuclear programme?". *Nature*, Vol. 321, p. 640, 12 June 1986. (Federal Republic of Germany)

NÉNOT, J. C., "Medical basis for the establishment of intervention levels". *British Journal of Radiology*, Vol. 60, pp. 1163–9, 1987.

ORGANISATION ECONOMIC CO-OPERATION & DEVELOPMENT, *Chernobyl and the Safety of Nuclear Reactors in the OECD Countries*, Paris: OECD, 1987.

RICH, V., "More compensation in Finland for nuclear accident victims". *Nature*, Vol. 325, p. 654, 19 February 1987. (Finland)

ROTBLAT, J., "A tale of two cities: Hiroshima & Nagasaki: a new look at the data". *New Scientist*, Vol. 117, pp. 46–50, 7 January 1988.

SCHEER, J., "How many Chernobyl fatalities?". *Nature*, Vol. 326, p. 449, 2 April 1987, *followed with reply by* FREMLIN, J. H., *Nature*, Vol. 327, p. 376, 4 June 1987, *followed with replies by* SCHEER, J., *Nature*, Vol. 329, pp. 589–90, 15 October 1987, *and by* FREMLIN, J. H., *Nature*, Vol. 329, p. 590, 15 October 1987.

SWINBANKS, D., "Chernobyl takes macaroni off Japan's menu". *Nature*, Vol. 329, p. 278, 24 September 1987. (Japan)

TRABALKA, J. R., EYMAN, L. D. & AUERBACH, S. I., "Analysis of the 1957–1958 Soviet nuclear accident". *Science*, Vol. 209, pp. 345–53, 1980.

TRICHOPOULOS, D., ZAVITSANOS, X., KOUTIS, C., DROGARI, P., PROUKAKIS, C. & PETRIDOU, E., "The victims of Chernobyl in Greece: induced abortions after the accident". *British Medical Journal*, Vol. 295, p. 1100, 31 October 1987.

WRIGHT, J. K., "Emergency planning". *British Journal of Radiology*, Vol. 60, pp. 1177–80, 1987.

Index

About the Author

Dr Richard Mould is an internationally known medical physicist, cancer statistician and radiation historian. As well as serving as a technical expert to the World Health Organisation and the International Atomic Energy Authority, he was a United Kingdom Government delegate to the IAEA Chernobyl Post-Accident Review meeting in Vienna, 25–29 August 1986. Dr Mould has written about 100 papers in learned journals and has sixteen books to his credit.